Nefasylum

© 2024
P.S. Winn

This book is a work of fiction, names, places, characters and happenings are a work of the author's imagination. Any resemblance to people, places or actual happenings is purely coincidental. All rights reserved. No part of this book may be reproduced or transmitted in any form, or by any means, electronic or mechanical. Including photocopying or recording, or by any information storage or retrieval system, without the authors permission.

Chapter 1

Sitting on the hard wooden bench, the old man's light blue eyes stared across the road.
Most days, rain, shine, and even snow, you could find him there. His family, those who had once talked to him, were now sitting on the other side. Others, he never saw, or spoke with.
He wondered how long until he stepped through that veil to join the ones who cared.
At least, he hoped that he would be heading through the veil to a better place, and not a worse one.

Watching him, from just up the block, darker blue eyes frowned. The much younger woman had seen the man every day on her walk home from the newspaper office where she worked.
Today was the first day she had seen tears in the man's eyes. That disturbed her.
Although the two had never talked for long, they always exchanged some type of greeting.
Stepping closer, Melissa Carlyle sighed with concern.
"Are you alright? I'm not being a busy body. You just look so sad today."
Melissa didn't mention the tears, still making his eyes glisten.

The man groaned.
"To be completely honest, I'm not sure how I'm feeling. My best friend just passed away. Poor Mason. He wasn't that old. They say he killed himself, but I know better. The guy was thirteen years younger than me. Damn, Mason never bothered anyone. He was only fifty five, He had a lot of living to do."

Sliding onto the bench, Melissa was doing some figuring in her head. If the man's friend was fifty five, then the man she was looking at, had to be sixty eight. Truth was, she had thought he was fifteen, or twenty years older than that.
As she sat, Melissa hoped the man wouldn't move away, thinking she was bothering. When he didn't, she grinned. The slight smile didn't last long.

Perhaps because the man was frowning.
Melissa was sure he would tell her to get lost.
The man looked at her. He pushed white hair back, away from his light blue eyes.
He knew he needed a haircut, but wasn't all that concerned about his looks.
"It's probably time we introduce ourselves. I mean we at least say hello everyday. In fact, I appreciate you being so nice, and talking to an old man. Not everyone does that. My name is Luke Harris."

Missy grinned.
"It's good to meet you Luke. I'm Melissa Carlyle, but you can call me Missy. Everyone does."
Reaching over, Missy shook Luke's hand, careful to do so gently. The large hand was twisted with what Missy had to guess was arthritis. The grip she received, was much stronger than she had thought it would be.

Luke grinned.
"I don't get a lot of young, beautiful women stopping by here, let alone talking with me."

Shrugging, Missy grinned. "Thanks for the compliment, Nice of you to say that. I see you here most days. Do you live close by?"

Luke shook his head. "Not really, I live about a half mile up the road."
He lifted a hand, motioning to the building, he watched each day.
"I'm here because of that place."

Seeing where Luke's finger pointed, Missy frowned, confused.
"You mean the old Nefasylum Building? That place closed years ago. I hate the name. What is that supposed to be? It sounds like a nefarious asylum. A place where bad things happen."

A grunt sounded, then Luke nodded.
"You're damn close. That place is an asylum. At one time it was run by Doctor Malcolm Neff. He named the asylum. It's actually pronounced nuh,fess,uh,lum, if you sound it out. I don't know who is in charge of the place now. I know they are responsible for Mason's death. Just as sure as the two of us are sitting here."

Missy frowned. "I'm twenty eight. I was told that thing closed a few years before I was born."
She pointed at the building sitting on the hill, across the way. The foliage had grown around the off white colored building, making the place look like it was part of the landscape.
"Look at that place. It's not open. I think it's on the verge of collapse. Who would be allowed to work in a run down building like that?"

Sighing, Luke turned to stare at Missy.
The light blue eyes still glistened with the tears, that were slowly drying up.
"That place is still open, and they killed my friend Mason Davies. He was a victim of their experiments, just like me. Only thing different between him and I, was after all this time, Mason just couldn't take the images in his mind anymore. Now he's gone, and I'd like to kill each of those people who pretended to be doctors."

Blue eyes, several shades darker than Luke's, frowned.
"I'm not sure I understand. What kinds of images?"

Luke shook his head of white hair.
He let out a long sigh.
"That's a long story, and not a pleasant one."

Blue eyes lit up as Missy smiled.
"Would you be willing to tell me some of it over dinner? I have a big casserole at my place that I could never possibly eat. My neighbor brought it to me. She's a good cook, but I think she must have worked at an army barracks at one time or another. She cooks for a whole fleet of people every time she makes a meal."

The frown on Luke's forehead made a deep crease.
"I would love to do that, but do you really think it's a good idea? I mean an old man like me, joining a pretty young thing like you for dinner. Some people would say you're asking for trouble with your invitation."

Laughing out loud, Missy shook her head.
"In my apartment building, everyone is nosy, and watches what the person next door is doing. A few are over protective of me, as well. If you make me nervous, I can always start screaming. Believe me, someone would come running. Will you please

come and have some food? I'd love to hear about that building you stare at every day."

Luke sighed. "Okay, you talked me into it. I'm too old to be much of a threat, anyway. I'm less sure about your hearing about that place. It was, and always will be evil."

Looking over at the building, thinking what the conversation was about, Missy felt a shiver run up her spine.
The place did look evil.
She waited for Luke to stand, then she motioned ahead of the two of them.
"My place is just a couple blocks away. Are you okay walking that far?"

Luke chuckled. "I'm old, but not decrepit. Lead the way, I'll be fine. I'm sixty eight, not eighty six. I actually walk about everywhere I go."

Nodding, Missy started walking. She was surprised to see Luke was moving just as fast as her. Then again, like he said, he was sixty eight, not eighty six. The man just looked older. The wrinkles in his face, to Missy, were like an old treasure map. Each wrinkle and fold, probably had an interesting story behind it. She loved stories.
Missy worked at the newspaper in town. She was a secretary now.

She had dreams of someday being a full fledged reporter. Her boss, Hank Roberts, kept telling her that she just needed to bide her time. Missy didn't think she had the patience for doing that.

The two approached the six unit apartment building. It was a brick building, that could use some work, but all in all, Missy was happy there. She turned to Luke.
"My apartment is around back, but it's on the first floor. I hate stairs, especially when it gets cold. I'm not a fan of clearing ice off of steps."

Luke nodded. "I hear you on that. My house is on the main level, but I have to clear my sidewalk, or hire someone to do it."

The two walked to the back of the apartment building. Going to her door, Missy kicked some leaves off the step, before she unlocked the door. She laughed.
"I'm not a big fan of cleaning leaves away, either."
She pushed the door open wide.
"Go on in."

Stepping in, Luke turned and waited for Missy. She was smiling.
"This is the front room, but we can head to the kitchen. Just follow me."

The two walked through the living room. Luke glanced around. He could see a sectional couch, and a recliner sat in the room. He liked that the place looked clean, and neat. He noticed several pictures that looked like they held newspaper clippings on the walls. From the living room, the two moved into a short hallway. Missy pointed to her right.
"The bathroom is just down there, if you need it."
Then she pointed the other direction.
"And the kitchen is this way."

Nodding, Luke followed Missy into the kitchen. The room the two stepped into, was much brighter than the living room, or the hallway.
A large, picture window looked out into what Luke thought was some kind of park. He could see several people sitting on benches, beneath tall trees. The leaves were mostly changed from there normal green color. The vibrant, orange, yellow, and red leaves looked even brighter as the sun hung low in the sky. It wouldn't be long until the sunset came, and changed the look of everything.

Missy pointed at her table.
"Grab a chair. I'll just heat up some of this casserole. Tell me what you like to drink. I have an assortment of things. Let's see, coffee, lemonade, punch and even some beer, or wine, if you'd rather have that."

Taking a seat, Luke grinned.
"I'd like a beer, but I better stick with coffee. If you really want to hear about Nefasylum, I should keep my head clear, for dredging up old memories. Plus, I need to walk home later."

Staring at Luke, Missy smiled. While she talked she made coffee, and heated the food.
"Of course I want to hear about that place. I have to admit, I heard a few rumors, and some stories, but never knew which to believe. I work down at the Mansfield Blotter. Some day, I'd love to be a reporter. So far, no one has asked me to write anything though. The paper's been around for a lot of years, In the archives there's several stories about that place. I bet none of them would tell me the things you could though."

Luke sighed. "A lot of the stories I have aren't very pretty. Nefasylum was never a good place. The man I told you who named that place, was evil. He ran things his way, and no one seemed to question the terrible things he did."

Missy's blue eyes widened, as she ran a hand through her long, wavy, dark hair.
She was about to reply, when the microwave dinged.
"Let me get the food, and grab us a cup of coffee. While we eat we can talk."

Moving to the counter, Missy reached up into the cupboard, grabbing plates and two mugs. She put servings of casserole on each plate, and placed them on the table. Then she grabbed the coffee.
Returning with the mugs, she frowned.
"What do you take in your coffee?"

Luke smiled. "Just black is good for me."

Smiling back, Missy laughed.
"I can't drink mine without sugar."
Missy headed back to the counter for the sugar bowl, and a spoon. Then she took her seat. She could see Luke hadn't started eating yet.
She smiled, and pointed at his plate.
"Go ahead, dig in. My neighbor, June Forsyth, she really is a great cook. I don't know what she calls this casserole, but everything she makes is wonderful."

Nodding, Luke took a bite, and smiled, nodding.
"Yep, its good alright. Tastes a bit like macaroni and cheese, but she's put several other things in this."

Missy nodded, but wasn't thinking about food. She took her first bite, chewed, then swallowed, before staring across the table at Luke.

"So, why do you think that asylum was responsible your friend's death?"

Luke sighed. "Maybe because the two of us spent time in that place. Mason was never the same, after the two of us finally got out of there."
Luke smiled. "You're not nervous because I spent time in an asylum are you?"

Missy shook her head. "Hell no. I don't even know what you would have to do in order to be placed in there. You're not a serial killer, or something equally sinister are you?"

Leaning his head back, Luke laughed heartily. "If I was, I'm sure I wouldn't tell you. But the truth is, I was just a young guy who made some bad choices. Nothing like a murder or anything close. When I got put in there, my biggest charges were doing the drugs I also sold. I was only twenty when a judge threw me in that place. No hard drugs like some are hooked on. I mostly sold marijuana, sometimes a little hash. Occasionally I was known to sell a few hits of speed. I was making some good money, and got careless. A lot of people in the asylum were in the place for committing similar crimes. You could find people in the building who did worse, but not many. Those who committed murderer and pleaded insanity. People of all ages were in that place. Many who were a lot younger than me. When Mason was brought in, he was only thirteen. By then, I had been in the asylum for six years."

Missy frowned. "How long were you up in that place?"

Luke grunted, his eyes narrowed.
"The asylum took ten years of my life. Ten damn, rough, long years."

Missy sighed. "And what about Mason? How long was he in that place?"

Lifting his fork, Luke took another bite of casserole, and ate it before speaking.
"Four years, but Mason went through a worse hell than most in those years."

Misty nodded. "You said your friend was dead because of that place. Why do you feel that way?"

A deep sigh sounded. Luke was quiet for a long time. Missy wondered if she'd asked too many questions. She was relieved when Luke shrugged, and spoke again.
"Because I knew what those bastards did to me, and the fact is, Mason received worse treatment than me. We were both released in nineteen ninety two. From the day Mason walked out of that building,

until a few days ago, when he died, he never had one night of sleep that wasn't filled with nightmares. Those people killed Mason, just as sure as I'm sitting here, they did."

A frown crossed Missy's forehead, leaving a faint line with the motion.
"I thought that building was a place where some kind of psychiatrists, psychologists, and other doctors were working to help people with their problems."

As a grunt sounded, Missy felt she would be hearing a lot of those as Luke told her whatever he was willing to share.
"Those people in charge never gave a damn about helping anyone. They were too busy thinking of the experimentation they could get away with. The truth is, they are still doing that. I heard Doctor Neff is gone from this world. I hope the hell that's true. I also hope he died the most painful death imaginable."

Seeing the worry in Missy's eyes, Luke laughed.
"Am I scaring you yet?"

Missy shook her head, and laughed a bit nervously.
"Not really. I think this doctor you are talking about would scare me more."

She frowned. "I would love to hear this whole story. Right from the beginning."

Taking the last bite from his plate, Luke took time to chew, swallow, then take a drink of his coffee, before he stared at Missy. Then he laughed. "That would take a hell of a lot of time. I don't think you want to hear an old man's rambling on and on, and for days. Look out your window. It's already getting dark outside. I should head for home."

Missy smiled. "Tell you what. If you agree to tell me the first part of your story tonight, I'll drive you home. My car is outside. Then, I'll feed you every night until you tell me the whole story."

The room was quiet long enough, Missy was anxiously thinking she had scared Luke away. But then he smiled, and nodded at her.
"I'll make you a different deal. I will take you up on the offer tonight. After that, we take turns making dinner, some at my place, some held here. One more thing, if you want me to do talking tonight, and agree to drive me home, I'll exchange this coffee, for that earlier offer of a beer."

Excited, and relieved, Missy smiled, and nodded. She stood from the table. Going to the refrigerator, she grabbed a beer. She placed it in front of Luke.

"Here you go. If you've had enough to eat, just let me move this stuff out of our way, then I'm all ears."

Missy cleaned off the table, and turned on the kitchen light. She hoped the warm glow would add some light to what she felt was going to be a dark story. Then she sat down, leaning her elbows on the table. She placed her chin on her folded hands and nodded.
"Okay, from the beginning."

Luke nodded, but took time to take a long drink of his beer, thinking of how to start.

Chapter 2

Luke cleared his throat, then began talking. "Like I said before, I got in trouble for selling, and also partaking of some drugs. I actually was just lazy, to tell you the truth. I didn't want a nine to five job, like others were getting. I found I could sell marijuana and make more money than my friends."
Luke laughed. "What's funny, is now that I have arthritis, my doctors actually have prescribed marijuana for me."
He shook his head.
"Life is ironic. Anyway, you want to hear this story."
He smiled. "If I start rambling on too much, just yell at me."

Missy smiled at the thought, but nodded.
"Don't think I won't."

Luke laughed. Then became somber.
"The big mistake I made, and I'm sorry for it now. That's the truth, I think about the mistakes I made, a lot. I should have never done them. I sold joints down at the local high school. I was twenty six at the time. I never thought too much back then about selling a joint or two, to a seventeen, or eighteen

year old. One of those kids was a damn undercover cop. I swear that bastard lied about his age to get into the police force. I'll believe that to my dying day."

Luke shook his head.

"Anyway, I went before the judge. They were talking about ten to fifteen years in jail. I was given a free defense attorney. This guy said I should plead guilty, but also plead insanity. You know, because of the drugs I was on. I was told I might get away with a year or two up at Nefasylum. A lot of people were doing that. Back then, no one actually knew what was going on up there. To me, a year or two in that place sounded a hell of a lot better than ten or fifteen in a cell. I agreed to make the plea. I had to take some tests, I pretended to be a little crazy, in order to seem insane."

Missy frowned. "But you were in that place ten years. Is that what you were sentenced to?"

Luke shook his head.
He ran a hand through white hair.
"No, I was given eighteen months in that place, with another eighteen months probation. All of that was contingent on what the doctors decided upon after further testing. My behavior, good or bad, also would be a factor. So, I became a patient of Doctor Malcolm Neff. He was the man to greet me when I first arrived."

Thinking of the man, Luke scowled. Missy knew he was reliving the moment in his mind.

"The guy had brown eyes, the color of mud. He was about five eight, and very thin. He was the nervous type. His nails were chewed down so far they were barely visible. My first thought, seeing him, was if someone should be a patient in that place, it should have been him. When he spoke to me, his voice was a deep baritone. It sounded strange coming from such a small guy. I still had on handcuffs when the guards transferred me to Nefasylum. Doctor Neff had them removed, then sent the guards on their way."

Luke laughed. "I should have gotten a hint about the guy, when the two guards nearly ran on their way out of the room."

He shrugged, and took a drink of his beer.

"Doctor Neff took me to a room, and closed the door. The moment that door slammed shut, he drew back a hand and slapped me across the face. I never expected that. I was in shock. Normally, I would have hauled off and slugged him back. Wiping at my cut lip, I stared at him, hoping he could see the hate in my eyes. The guy laughed. He told me if I made any wrong moves, he had a way to call in his guards. I guess the asylum had a pager system set up. That was in nineteen eighty two."

Missy nodded. "I think they had pagers back then. You know, especially in hospitals, and things like that. Why would that guy just hit you like that? It doesn't sound like you did anything to him."

A grunt sounded, as Luke shook his head.
"Who knows why he did anything he did, or ordered others to do for him. My being in that place was excuse enough for Neff."
Luke shrugged.
"The first couple of weeks in that place weren't too bad. I got there in spring time. The residents were allowed to go outside if we wanted. Back then, most of us smoked. I quit, after I was finally out of that place, but couldn't manage to break the habit, while I was inside. It was close to summer, when I knew things were going to get bad."
Luke drew in a deep breath.
"A couple of guys came to my room. They said the doctor wanted me upstairs. He called the place his consultation room. I had never been up there. None of the people I talked to told me anything about it. Not that any one in the place was close to anyone else. We kind of stayed to ourselves. I was taken to a large room that had several chairs in it. They looked kind of like something you'd find in a dentist office. I was seated, then my arms and legs were strapped down. I struggle, and tried breaking loose. No such luck. The guards left when the doctor came in. Doctor Neff stared at me, then smiled. That grin never touched those murky brown

eyes. He reached into a table drawer and pulled out a needle. That damn thing looked huge to me."

Light blue eyes opened wide, like Luke was still seeing the needle. Watching the man, Missy could imagine he actually could see that terrifying image in his mind. Even after all these years.
Her heart broke for the man. She didn't say anything, waiting for Luke to continue.
After a few seconds, he did.

"The needle plunged into the inside of my elbow. I could feel the stinging first, then a cold sensation. A few moments later, the room was spinning. The doctor's face went in and out of focus. He was holding up a metal spoon, asking me to bend it. I struggled against my restraints. Thinking, he must want me to use my hands, but he was shaking his head, telling me to use my mind. He repeated that order over and over."

As Luke shook his head, Missy frowned.
"What the hell? Why would he think you could do that?"

Luke sighed. "I think he gave me a shot of something like LSD. I know experiments were done with a variety of mind altering drugs, and the capabilities when they were used."

Missy shrugged. "I've heard a few conspiracy theories on that. Watched a couple of strange movies, but never thought they were actually true. Are you talking about something like what the government did with like that stuff they called the MK Ultra Project? Was that even real?"

Luke nodded. "Oh it was real, and our government has done that and much worse."

Missy frowned. "Here in Mansfield?"

Again, Luke sighed, and nodded.
"Yes, even here in the tiny town of Mansfield. I think choosing a small town like this to build the Nefasylum, was one of the reasons it was done. No one expected that. Doctor Neff was given a big budget, and a lot of people who were willing to look the other way when he was doing his strange experiments. You have to remember that our government is always looking for ways to build bigger, better and more unique weapons for their military."

A groan sounded. Missy was shaking her head. The long brown hair swayed. The dark blue eyes drooped with her frown.
"I told you I work at the paper. My boss might want to hear what you went through."

Luke gave Missy a sarcastic smile, and shook his head.
"I don't think he'd believe me. Maybe you won't when you hear the other stories I have rolling around in my head. You might just think I'm an old man with an overactive imagination."

Missy shook his head. "No, I've seen the way you stare up at that building. I can hear the truth in your voice when you're talking about the place."
Reaching across the table, Missy placed her hand on Luke's.
"I believe the little you've shared, and have a feeling nothing you tell me will change that. You might scare the hell out of me, but I would never doubt what you are saying."

Luke smiled.
"I appreciate you saying that."
He nodded, took a drink of his beer, and grinned.
"Okay then. I'll just finish with my first installment, then call it a night."
He sighed.
"Over the next few days, I was given those shots every few hours. I don't think that shit had time to leave my system, before the next injection was given. By the fourth day, I was higher than a kite, hell, I was higher than a space ship. I climbed out of a window on the top floor of the building. I thought I could fly. Luckily, one of the psychiatric

techs saw me, and pulled me to safety. I wondered if Doctor Neff would have enjoyed seeing me jump. Just a casualty of his experiment. That would have shown they could give the drug to an enemy and had them jumping out windows."

Looking across the table, Luke watched Missy's face. He had heard her declare she would believe anything he said, but saying something and doing it, was not always the same thing. He'd lived long enough to learn that lesson in some harsh ways. A lot of his own family, and too many of his friends had written him off as crazy.

Watching Missy, so far, he thankfully didn't see any doubtfulness. Luke continued.

"I was taken back to my room, and allowed some much needed rest. Later, I talked to a few others who were stuck in that place, same as me. A few of them had been given those shots as well. None of them had tried to fly, like I did, but one had committed suicide."

Missy gasped. "Oh no, how awful. Why wasn't that death investigated? Didn't that person have family asking questions?"

Luke shook his head, and grunted.
"Don't forget, that place was a mental facility. We were all listed as crazy, and already a bit unstable. A suicide wasn't something that raised eyebrows. When people did ask questions, they were paid an

insurance settlement for the death of their family member. I'm sure they were also asked to sigh some type of non-disclosure agreement. Doctor Neff had a way of talking his way out of trouble. That guy had the Government behind him. Hard for anyone to fight against that kind of power."

Missy nodded at what Luke was saying, but her mind was ticking away. She knew that if someone could get a story like she was just beginning to hear from Luke, out where the public could learn about it, a lot of things could happen. The world today, with social media being what it was, could sway opinion in ways it couldn't before.
If Luke's thinking about Nefasylum still being operational was accurate, she needed proof of that. Missy wanted to be a reporter. What better way to do that, than to break a story of epic proportions wide open? If she could also help heal an old man's wounds and give Luke some kind of closure at the same time, even better.

Staring at Missy, Luke could almost see the wheels turning in her mind. He didn't know what she was thinking about. Right now, all he knew was sharing a bit of his story, along with drinking a beer, was making him tired. He lifted the bottle, and drained the last drop of the liquid, then placed the empty bottle on the table. He grinned.

"I feel like I talked enough for the first night. If you're still willing to give me that ride, I'd like to head home. Tomorrow, I want to cook the dinner, though."

Nodding, Missy stood.
"Great, I get off at five. Would that be too soon to head to your place?"

Laughing, Luke shook his head.
"Nope. Five it is."

The two took time to exchange phone numbers, before Missy drove Luke home, and they separated for the night.

Chapter 3

    Looking at the clock, Luke smiled. It was almost time for Missy to arrive. He had dinner ready. He had put everything in his oven to stay warm while he waited.

Outside the house, Missy was just driving into Luke's driveway. She had dropped him off the night before. The night had been dark enough she hadn't gotten a good look at the house.
Now, she stared at the place. The cabin type home was tucked away behind large trees. The brilliantly colored leaves were just beginning to fall, but enough remained on the branches that not a lot of the home was visible. She hoped that Luke would be willing to give her a tour of not only his home, but the surrounding property. Living in an apartment, Missy envied anyone who had an actual house.

Stepping from her car, she grabbed her purse, and walked up a path littered with a few leaves to the front door, and knocked.
She waited a few brief moments. She smiled when Luke came to the door.

He was also smiling.

"Glad you could make it. Come on in. Sorry I didn't get time to sweep those leaves away. I did it earlier, but it's a never ending job this time of year."

Missy smiled. "Actually, I like to listen to that crunching sound. The colors are amazing as well. Mother Nature knows her business."

Luke laughed. "She does, but she has a wicked sense of humor sometimes."

The two stepped into Luke's living room. Outside, the air had been slightly chilly, but the room was warm and cozy. Missy saw a wood burning stove in the corner of the room. She pointed at it.
"Looks like you're ready for cooler weather."

Looking at the stove, Luke nodded.
"I hope so. I used to go out and cut my own wood supply, but I can't do that any more. I know a couple of younger guys who go up in the woods. They give me a good deal on the wood I need. I miss the trips though. I like being outdoors. I feel at peace there."

Glancing around the room, Missy noticed several large pictures, mostly of landscapes hanging.
"Those are beautiful. Who's the artist?"

Luke smiled.
"At least one of them was done by me. A few others were done by people who spent time up at Nefasylum. I've had all of those paintings for a lot of years. I can't part with them. I wouldn't sell any of them for all the money in the world."

Moving closer, Missy stared at a mountain scene. She looked at the name scrawled in the corner and smiled seeing it said L. Harris.
She pointed at the signature.
"Is that you?"

Luke nodded. "It is. My favorite painting is the one sitting next to it though."

Turning her head slightly, Missy stared at a painting of a desert scene. The name at the bottom was hard to make out. Her eyes narrowed as she leaned closer. She read it out loud.
"Mason Davies."
She turned to look at Luke.
"Is that your friend Mason?"

A sad look came into the light blue eyes, as Luke nodded.
"It is, or it was. Mason was from a little town down in the southwest, near what they call the four corners area. That's where Colorado, Arizona, Utah, and New Mexico meet. I've never been there,

and Mason always said he couldn't wait to get out of there. He had a rough upbringing."
Luke sighed.
"Let's go out in the kitchen. I can tell you more about it, and we can have our dinner."

Missy nodded.
"Sounds good. When we're done, will you show me your house, and your yard? You are so lucky to have a house instead of a little apartment."

Luke shrugged.
"Not much to see, but yeah, I will."

The two walked from the living room, through a small room that Missy thought might have once been a dining area, but now was more of a storage area.

In the kitchen, Luke pointed at table with four chairs. The table was a round, wooden table. Instead of legs, it had a large pedestal running down the center.
"Take a seat, everything is ready. I have coffee, hot chocolate, or some wine. What would you like to drink?"

Nodding, Missy smiled.
"I'm driving. Some hot chocolate would be great."

Luke nodded. "Okay. Let me grab everything. I can tell you about Mason while I'm doing that."

As he began working, pulling things from the oven, Luke talked.

"When Mason was born, his mother died. Some complications with childbirth. I have a feeling in the area Mason was from, that was common. What family he had was poor. I think they lived on a Native American reservation or something. He didn't talk about it much. Anyway, his dad was an alcoholic, and when Mason's mother died, Mason was given up for adoption."

Luke had all the food on the table, then grabbed the drinks. He pointed at the table top.

"Go ahead and start eating. You fed me your neighbor's casserole, so in turn, I made one of mine. This is shepherd's pie. Hope you like it. If not, I have leftovers for about a week."

Missy laughed. "Shepherd's Pie is actually one of my favorite's. Thanks for making this. So what about Mason?"

Sitting down, Luke filled his plate. He took a few bites before he spoke again.

"Instead of being adopted, Mason spent his time in an orphanage. I guess he got in a lot of trouble. Nothing big. Not at first anyway. But from what he spoke about, Mason began getting thrown in juvenile detention when he was just in third grade."

Shaking her head, Missy sighed.
"That's terrible. Our crappy system is to blame for that. As a society, we should be doing better. How can we expect to have productive citizens when the system fails them?"

Luke nodded. "Yes, I have learned that lesson in my own life, and see it in many others."

Missy nodded. "Before you share any more of your story, I need to tell you how good this food is. Thanks for cooking tonight."

Luke smiled. "My pleasure, and thanks. I like to cook, but with just me eating, I don't do it often. I used to cook for Mason. He lived here with me, until he died."

Staring at Luke, Missy frowned.
"I didn't know that. How long did the two of you live together? Did you buy this house together?"

Luke laughed. "The truth is, neither of us bought this house. I inherited it from my grandma Spencer. She was the one person who always believed in me. Grandma Spencer was my grandma on my mom's side of the family. She left this house to me. I moved in when I was released from Nefasylum. She passed away the year before that. When Mason was released from that place, he was only seventeen, and went to a half way house for six months. As soon as he turned eighteen, I had him move in here. Mason never was the person he had the potential to be, after Dr, Neff got through with him. Still, he was able to hold a job, and deal with things for almost four decades. The nightmares, and the horrors he lived through, even in waking hours finally became more than he could handle. Mason shot himself, but it was really Doctor Neff's finger that was on the trigger."

Missy frowned. "Did Mason commit suicide here in the house?"

Luke shook his head. "No, he went up to Falcon Lake. Mason liked to go up there and fish. He didn't even eat fish, but it was his way of relaxing. He left here one morning. He said he would be home shortly after lunch time. He didn't show up by late afternoon. I had a neighbor drive me up there. We found Mason's old truck. A short distance away we found his body. I wish he would have told me things were getting bad enough that he couldn't take life anymore. I don't know if I could have helped him, or found someone who could have done that. Mason was adamant about not going to any kind of psychiatrist or psychologist. Doctor Neff did that to him as well." Luke let out a sigh.

"Malcolm Neff ruined a lot of people. I heard a few years back the man was found murdered. I don't know if that's true or not. I hope it is. I hate saying that, but I can't find a way in my heart to forgive what that man did. I suppose it wasn't just him. Others had to know what he was doing. He was given the money to continue his experiments. There had to be some type of monitoring what was happening."

Looking at Luke, Missy made a mental note to check into the doctor's death. Something like that should be easy to verify.
"You were in that place for ten years. How long was Mason in there?"

Luke sighed. "Mason was in there four years. He arrived when he was only thirteen. When he was brought in, I had already been there six years."

Missy's mouth dropped open.
"Oh hell, I still don't understand how you ended up in that place for so long."

Seeing Missy's plate was empty, Luke pointed at it. "Are you full?"

Missy nodded. "I am, and again, thanks, it was delicious."

Smiling, Luke nodded.
"Good. If you want to, before I continue talking, and before it gets dark, I'll show you around the place. We can go outside first."

Nodding, Missy smiled.
"I'd like that. I could do the dishes first though."

Luke laughed, shaking his head.
"No, but thanks. I'm old school, but I have a dish washer that can take care of those dishes. C'mon, let's go out back."

Standing, Missy followed Luke from the kitchen, through a laundry room to the back door.
The two stepped out into a large back yard.
Luke motioned around the area.
"The yard around this house is just over a half an acre. The whole area is fenced in."
He pointed to the south in the back of the yard.
"I have a garden out there, and a couple of fruit trees. Two cherry, and four apple. Along that fence to the east are raspberry bushes. If you head over to the west side there's a small creek running along the side. I have a pump in the water I use for watering the area."

Missy stared wide eyed at Luke.
"What a great set up."

Luke shrugged. "My Grandma and Grandma Spencer get the credit for that. I do some weeding, and the watering, but they planted all of this years ago."
He smiled.
"The trees out front are mostly for shade. I built that picnic table and put the covering up over it, but not much else. When you visit next time, this place could have leaves, or snow. Around this area, you never know what to expect."

Not born and raised in Mansfield, but after spending a couple years there, Missy nodded. "That's true, I've seen it begin snowing in early September and seen the last snow in June around here."

Luke smiled. "Okay, let's go in. We can walk around the house, then go back in the kitchen."

Going back in the door, they had come out of, the two headed back through the laundry room, then the kitchen. From there, Luke stepped into the living room. Off that room, a hallway lead down to three bedrooms, and two bathrooms. Luke's room and another, that had been Mason's, were similar in size. The third bedroom was smaller, and like the room Missy thought might be a dining room, was used mostly for storage. The bathrooms were separate from the bedrooms.

Missy asked to use one, before the two headed back to the kitchen.

Once the two arrived back in the room, Missy helped Luke clean up, then load the dishwasher, before the two sat down at the table with mugs of cocoa.

Luke stared at Missy.
"So, were you born in Mansfield? You said you knew about the harsh weather here."

Missy nodded. "No, I wasn't born here. My dad was from a town not far from here. My mom was from back east. They met when my dad went to college back there."

Luke nodded. "And now you work at the paper. Are you a reporter?"

Missy grunted. "Hardly, but someday I'd like to be. I went to college so I could be a journalist. My boss, Hank Roberts, he keeps telling me, that I have to be patient. Our paper is small, just like the town, and there aren't openings for a journalist right now."

Luke laughed. "Hey. Maybe you just need to give your boss the right story. If you could show what happens up at Nefasylum, prove the past atrocities are still happening, he'd have to print your story, don't you think?"

Missy frowned. She had thought about that before. She'd first been so worried about Luke, then had been fascinated by what he was sharing. She had given some thought about the whole thing being a big story. her newspaper might be interested in. Reminded of that now, her eyebrows lifted with interest.
"You might be right. All that you know might be hard to prove though. Especially the part about something still going on up at that place. I mean it looks like a building waiting to disintegrate in even a light wind."

Eyes narrowed, Luke stared at Missy.
"Looks can be deceiving. The two of us should take a ride up there. I think I can prove to you what I'm saying is true. Not the part about the past, but the part about the present."

Now, Missy stared at Luke. Her head tipped slightly to one side. Dark blue eyes stared at lighter blue ones, with growing intrigue.
"If you mean it. Then let's go. I'll drive."

A few minutes later, the two were headed up the road. Missy was amazed she was actually driving to Nefasylum. The whole thing, to her, felt like a wild

goose chase. Then again, she believed the things Luke was saying about the horrid past of the building, why not the present?
She turned her head, hearing Luke talking.

He was pointing out the windshield.
"Go, and park under that large willow tree."

Nodding, Missy drove her car under the shadow of the tree. She turned off the engine, then turned to Luke.
"Okay, now what? The front of that building is a couple blocks away."

Luke grinned. "The front of the building is, but the basement of that place actually runs backward, right under where this car is sitting. The concrete foundation is larger than the main building."
Opening the door, Luke smiled.
"Just follow me. No one will know what we're doing."

Hoping the man was right, Missy got out of the car, and walked a step behind Luke. She was surprised how spry the man seemed tonight. If she had gone through even the few things Luke had told her about, she wondered if she would go any where near the building. Let alone at the pace Luke was going tonight.

He moved so he was squatting by a bush, and motioned for Missy to join him.
When she did, he whispered, and pointed toward a basement window, where a faint light could be seen.
"There's someone in there."

A frown made the corners of Missy's eyes turn down. She shook her head. Not because she didn't believe Luke, but because she didn't believe what she was seeing in front of her own eyes.
"What the hell? This place is ready to fall down. Why would someone be inside of that decrepit old building?"

Luke sighed. "Whatever they are doing in there, you can bet it isn't for the good of mankind. Come on, let's go take a look in that window."

Staring at Luke, Missy wasn't sure she had heard him right. The determined look on his face said more than any words could.
The two half crawled to the edge of the window, and looked in. The paint on the sill was peeling. The window itself was so dirty, it was hard to see anything. The two could make out the image of a man standing by a table, leaning on crooked legs.
Missy frowned again. She was feeling like she was going to have a permanent frown during this ordeal. Her voice was barely a whisper.

"Do you know who that is?"

The reply was also spoken quietly.
"No, any one working in this place wouldn't be someone from my past dealings with Nefasylum. I have to say the darkness in their hearts would look the same though."

Missy sighed. "We should go before someone sees us. I'm sorry I doubted you about this place still being used."

Luke grinned. "If you didn't doubt me, then I would have thought you were crazy. Never believe a wild story until you check the facts. A good reporter, like I know you will be, knows that."

Nodding, Missy also grinned. The two made their way back, crawling at first, then walking, when they felt they weren't in danger of being spotted. Missy felt relief flood through her, at the sight of her car. The two got inside and sat down.
Missy didn't start the engine right away. Instead she turned to stare at Luke.
"Where do we go from here? What would you like to happen?"

Luke sighed. "I don't know. I guess I'd like the truth to come out. Some day, I'd like to sit on the bench I always do, and watch that building torn to the ground. The atrocities that happened there, and I'm sure are still happening, need to be stopped. Shutting this one place down, might not stop much, but then again, it might raise awareness. Many people think that the atrocious things that happened decades ago in the insane asylums, stopped back then, when there was some reform. That's not the case. People need to know that."

Missy nodded. "You're right about that. I think we should call it a night. Should I cook dinner tomorrow?"

Luke shrugged. "If you want to. I can walk over to your place, but you'll have to drive me home. I do have an old truck, I let Mason drive. I don't have my license. It expired, and I never bothered to get it back. I don't usually need to drive anywhere. I walk about everywhere I need to go. In fact, I could even walk home after we have our talk tomorrow."

Missy shook her head.
"Nope, I'll drive you. I don't mind doing that. I should do more walking, but besides my walk to and from work, I drive everywhere I go. I feel lazy compared to you."

Luke smiled. "I doubt if you're lazy. Walking is a great way for me to clear my mind."

Missy had started the car, and was driving back to Luke's house. She pulled in to his place, and turned to face him.
"Dinner was great, and the rest of the night interesting. Thanks for showing me your house as well. It's a great place."

Luke opened his door, smiling.
"And you are a nice person. I'll see you tomorrow Missy. What time?"

Missy thought about it, then grinned.
"I get off at four tomorrow. If you're going to be sitting on that bench, you can walk with me after I leave work. We can go to to my house, and help get the meal ready. We can be eating by five."

Luke laughed. "Deal."

Chapter 4

The next day, Missy headed to work, thinking about the night before. She wanted to talk to Hank Roberts. The owner of the Mansfield Blotter was twice Missy's age. Hank was like a father to Missy. He had hired her as a secretary, knowing she wanted to be a journalist. The newspaper was a small town paper, with just a few employees. Hank didn't have an opening for a reporter, but hoped Missy would stay working at the place until he could move her into the position she wanted. Missy was thinking that Hank was only a year older than Luke's friend Mason had been. Luke has said Mason was fifty five. She knew Hank was fifty six. She wondered if the two had known each other. The possibility, though not likely, was there. Hank was from Mansfield, though he had spent a lot of years away from his home town.
Stepping into the office, Missy grabbed a cup of coffee, then stuck her head into, the always open door of, Hank's office.
"Morning Hank. Do you got a minute?"

Brown eyes frowned, but Hank nodded.
"Sure, come on in and sit down. Is something wrong?"

Taking a seat, Missy sighed.
"Nothing's wrong with me, but something might be wrong. Or maybe I should say something is off kilter. Do you know much about Nefasylum?"

Hank shrugged, brushing back his sandy colored hair, now getting a few white streaks.
"I know a few things. Does this conversation have something to do with that suicide that happened? The death of Mason Davies? I know he spent time in that place."

Staring at her boss, Missy was surprised by the accuracy of his question. She nodded.
"I guess it does. I see this guy every day when I walk home from work. His name is Luke Harris. Mason was his friend. The two of them were patients in Nefasylum."
Missy frowned.
"Did you know the terrible things that happened in that place?"

Hank shrugged. "I've heard rumors. Hell, this paper has run some stories on that place. That was before I owned it. Nefasylum closed back in two thousand. I bought the paper four years after that. You can search the archives for some of the stories. The place was run by a guy named Malcolm Neff. He was run out of town when some of his experiments were looked into. He was doing things that were outlawed back in fifties, and sixties."

Missy sighed and nodded.
"That's what Luke's been telling me. That place isn't completely closed Hank. I saw someone in the basement of that building last night."

Hank stared at her.
"What the hell were you doing up in that building?"

Missy grunted. "I wasn't in the building. I was on the outside looking in. I saw a light, and got curious. I peeked in the basement window, and saw a guy in there."

Now, Hank was frowning.
"No one should be in there. I think that building was condemned. It should be torn down. Maybe you just saw a homeless person, or even a drug addict, hiding out."

Shaking her head, Missy grimaced.
"That's not what I saw Hank. Luke told me that place was still open. I believe him. Someone, maybe even our own government, is doing something up at Nefasylum. Probably something they shouldn't be doing. If what they were doing was legal, and moral, they wouldn't be hiding out in the basement of a condemned building."

Staring at Missy, Hank frowned.
"How come you know so much about this Luke guy? Are you sure he's not dangerous?"

Missy smiled, and nodded.
"He's not some kind of desperate character Hank. He's a man who sits on a bench every afternoon, staring at an awful place where, he was, for the most part, a prisoner. Luke's friend killed himself. Luke says that was because of what that place, and Doctor Neff did to him. I believe that's true. I just want to learn more about the experiments that were going on. I want to compare them to what I am learning from Luke."

Hank sighed. "I can help you find the stories that were in this paper. I still worry about you talking with this Luke guy."

Missy laughed. "You're going to be super nervous then, he and I are going to be having dinner every night."

Hank stared wide eyed at Missy.
"What? You're kidding right? What in the hell are you doing Missy? You could get hurt."

Smiling, Missy shook her head.
"I don't think so. Listen, let me call Luke and see if

he minds you coming to dinner tonight with us. You can meet Luke. The three of us can talk about Nefasylum, and the truth of what happened. I'd love to have this paper tell that story."

Staring at Missy, Hank was thinking of what it would mean for the paper to break open a story like what Missy was telling him. He didn't actually believe someone was doing strange experiments up at Nefasylum, but it wouldn't hurt to hear what Luke Harris had to share. Hank nodded.
"Okay, I'd love to come to dinner, and have a talk with the two of you."

Missy smiled. "Good, I'll call Luke."
She hurried out to her desk.

A few minutes later, Missy stepped back into Hanks' office.
"Sounds like the three of us will be having dinner. Can you come by my place around five?"

Hank nodded. "I'll be there. I wouldn't miss this whole ordeal for anything in the world."

Going back to her desk, Missy divided her time between doing her work, and searching the newspaper's archives. She was thankful they had been digitized and she wasn't digging through a back room of physical papers.

She stared in horror at a few of the headlines. Missy knew the asylum had been opened in nineteen sixty five, but apparently it had taken years before the cruel deeds were taken notice of. It wasn't until the late nineties, that she saw the first story. The heading was horrific.

"Unwitting Patients at Nefasylum Sterilized."

Reading the article, Missy felt like she was going to throw up. She hadn't known that the asylum had held males, and females. According to the story, the atrocities were committed by the doctor on both sexes. Several people had come forward to say that they had been sterilized without consent. Missy was shaken her head, Who would do that and why? Another headline was equally hideous.

"Three patients die after months of isolation at Nefasylum."

Missy stared at the words in disbelief. The patients were found in barren rooms, without windows. They were malnourished, dirty, and in ill-fitting clothing, when they had been discovered. One of the attendants had reported the deaths. Doctor Neff had been questioned, but hadn't been found guilty of the terrible acts. Missy knew that meant the doctor had to have friends in high places. That infuriated her even more. What kinds of people allowed things like that to happen? Many deaths at Nefasylum could have been preventable.

Intervention should have happened before the first

person had died. It should have happened before one person suffered. Luke had said Mason had been put in that place when he was only thirteen. Missy wondered how many of the patients at the asylum were children? She felt like crying. She also felt so angry that if Doctor Neff was in the room, she would probably strangle the man.

Looking up at the clock, Missy shut down her computer. She needed to go meet Luke. She also was sick to her stomach after reading the few stories she had about Nefasylum.

Heading to Hank's office, Missy told the man she was heading out, and made sure he would be coming by for dinner. Then she left the newspaper office.

She walked to the bench where she saw Luke sitting. She waved at the man.

"Luke, hi. Are you ready to head to my place?"

Luke nodded. "Are you sure you still want me to do that?"

Missy laughed. "Of course I am. What a strange question. Get up, let's go."

Standing, Luke walked with Missy to her apartment. He was frowning.

"Does your boss think you're crazy for talking with me? Is that why he's coming to dinner?"

Frowning, Missy shook her head.
"No, why would he? I think Hank just wants to learn more about what happened. Like me, he believes if we forget the horrors of our past, we tend to repeat them."

Luke nodded, although he thought many people, hearing his story, felt he was crazy.
The two walked inside Missy's apartment. Luke pointed at the pictures on Missy's walls. He had meant to ask about them before.
"Did you write those articles you have framed?"

Shaking her head, Missy laughed.
"No, I wish I did. I want to be a journalist. Most of those were written by my boss, Hank. Two of them were written by Marie Parker. She was the top reporter at the paper. She went on to bigger, and better things. Now, Hank does most of the writing. We have one other reporter, George Benson. He's almost sixty five, and I hope retires soon. I would love to take his place."
Smiling, Missy shrugged.
"All good things, in time. Come on out into the kitchen. I have to fry up some hamburger. I though we could have make your own tacos for dinner."

Luke smiled. "Sounds good to me. What can I do to help? I'm fairly handy in the kitchen."

Missy shook her head. "You can sit down and talk to me. Everything is prepared. Well, except for frying the hamburger and throwing some tater tots in the oven."

Taking a seat, Luke sighed.
"I feel a bit worthless just sitting here."

At the stove, Missy had already began with her cooking.
"You're my guest. And you could never be worthless. Tell me how you knew that people were up in Nefasylum."

Luke sighed. "It was Mason who noticed it first. He never slept all that well. He liked to walk when he couldn't get to sleep. Sometimes he would wake up in the middle of the night, and go walking. He saw the lights in the basement up there. The two of us went back several times to look. We never got a good view of what was going on inside, but we could see the basement area, near the window, looked like a laboratory. We knew that the things we had gone through up there, hadn't ended when that place was shut down."

Nodding, Missy turned the meat to simmer, put the tater tots in the oven, then took a seat at the table.
"I can't even begin to imagine what is happening. Do you want a drink? I should put on a pot of

coffee, and maybe some hot chocolate for dinner. I have cold drinks in the fridge, but it isn't all that warm outside."

Luke nodded. "I'd like some coffee. You're right about the weather. I think we may be in for a cold winter. Fall came in warm, but has cooled off fast."

After Missy made coffee, and cocoa. She and Luke decided to have coffee for now. Missy got them each a cup, then sat back at the table. She had just gotten seated, when the doorbell rang.
"That'll be Hank. Sit right there, I'll go let him in, and bring him out here."

A few minutes later, Missy returned with Hank. She smiled at Luke.
"Luke Harris, I'd like you to meet my boss, Hank Roberts."
She turned to look at Hank.
"And vice-versa."

Luke stood, and shook hands with Hank.
"It's good to meet you. Missy talks highly of you."

Hank smiled. "Good to know. I'm sorry about your friend Mason. Such a tragedy. Missy tells me the two of you were up in Nefasylum together."

When Luke began to nod, Missy spoke before he could answer.

"Before we do any talking about that horrid place, I want to get some food in my stomach. After we talk, I know I won't have as good of an appetite. You two sit down, let me set everything out. I'll put all the ingredients on the counter, you can both make your own tacos. Those tater tots should be ready by now."

Walking over, Missy opened the oven and smiled. "Yep, nice and brown."

She pulled out the sheet that held the potatoes, and placed it on the stove. Then she grabbed all the ingredients from the fridge, and placed them on the counter. Missy set plates on the table, and spoke to the men.

"Grab your plates, get a taco shell, and fill it up. Make as many as you want. I have enough fixings for a room full of people. Anything we don't eat is going into a big bowl, and I'll have taco salad, probably for days."

Everyone made their tacos, then sat at the table. They talked while they ate, but since Missy asked, they didn't speak about anything related to Nefasylum. Instead they talked more about how, and why they were in Mansfield. Both Luke, and Hank had been born there. Luke had spent his whole life in the area. Hank had worked in several places as a journalist before returning and buying

the Mansfield blotter. Missy came from another part of the state. She had lots of family back there, but preferred being out on her own. She had laughed when she shared that information.
"Back home, I have two brothers, two sisters, my parents, and tons of other relatives. Being in a close family is nice, but it also means everyone wants to know every move you make."
Missy pushed her plate to the side.
"I've had my fill of tacos. I left room for a turnover, though. I hope you both did. I bought both apple and cherry down at the bakery."

Hank smiled. "I'll have cherry."

Luke nodded. "Cherry sounds good to me."

Grabbing the turnovers, Missy also opted for cherry. Once dessert was done, the three decided to have cocoa. Then they talking began in earnest.

Hank started the conversation, staring at Luke.
"In my time in Mansfield, I've heard a lot of horror stories about Nefasylum. I never knew how much was true. I've read several articles written by the paper I own, and by others."

Luke grunted. "The stories that don't make it into the paper, are the ones that hold true, sadly."

Hank frowned. "Did you and Mason spend time in that place together?"

Luke nodded. "We did. Mason was there during my last four years at the place. I was there ten years altogether."

Disbelief, and outrage, were clearly visible on Hank's face.
"Ten years? Damn what happened to send you to that place?"

Luke sighed. "I was taking, and selling drugs. Mostly, I feel that Malcolm Neff was looking for patients to experiment on. It didn't help that my lawyer thought pleading temporary insanity was a good thing. It was Neff who kept extending the time I was in that place. Mason though, hell he was just a kid. He was only thirteen the day he was ushered into that insane asylum. Neff didn't like Mason from the very start. Or, maybe he just felt that Mason, who had no family, no money, and was Native American, deserved to be used for his experiments. I really can't say. I just know from day one, Neff had it in for Mason."

Hank shook his head, and groaned.
"Being poor, alone, and Native American, aren't any kind of reasons. In my research, I heard about some horrific things. That doctor used outdated,

and outlawed practices. Things like sleep deprivation, electric shock, and even sterilization were listed."

Missy gasped. "I read about some of those awful things. How could that happen? That place was open until two thousand. Those things you're talking about were supposed to have been stopped in the sixties."

Luke grunted. "Not according to Neff. He not only did those things, to males and females, my friend Mason was castrated."

A shocked yell came from Missy, she covered her mouth. She wasn't sure if she was going to cry, or throw up. It took her a moment to get her emotions under control.
"I didn't know until I read some articles, that there were women in Nefasylum. How dreadful, and poor Mason."

Luke nodded. "It never should have happened. Mason never was able to lead a normal life. His mind was twisted by the things Neff did. As for women being there, the men outnumbered them, but several were there. I'd say for each woman there were four men. The females had rooms on the first floor. The males were on the second. We weren't allowed to talk with each other. Even our meal times were separate."

Hank frowned. "How many patients at a time?"

Luke thought a moment. "Of those who were more permanent, I'd say twenty five men, and about six women. The asylum also had patients who were only there for a few days. Those receiving mental evaluations, and things like that."

Missy frowned. "This is something people should know more about. Who owns that building now? If they are doing experiments again, that place should be bulldozed to the ground."

Luke shook his head.
"I don't know who owns it. It was a government run facility. I just figured when they closed the doors to that place, the government sold it. Now, I wonder?"

Hank's brown eyes narrowed.
"I have a few connections. I think I can find out who owns it. Hell, I might even be able to get inside."

Missy stared at him, wide eyed.
"If you can get inside that place, I want to go with you."

Luke nodded. "So do I. Through my life, I always thought I would never say that. I want to see what's happening, and try to stop any more atrocities from going on. That place screwed with too many people's minds. I don't know how many died, either in that place, or from the after effects, but it can't be allowed to happen again."

Both Missy, and Hank nodded. Hank looked at the other two at the table.
"If you can give me a day or two, so I can check things out, I'll let Missy know what I find."
He turned to Luke.
"Could I get your number. I'll give you mine. If you have anything to share, or just need to talk, you can give me a call."

Luke nodded. "I appreciate that. It's an amazing offer. You don't even know me."

Hank smiled. "Maybe not, but Missy trusts you, and that's good enough for me."
Then he sighed. "I could stay and listen to Luke all night, but I should get going. I have work to do."

Luke nodded. "I should get going myself."
He turned to Missy.
"Are you still alright giving me a ride?"

Hank frowned looking from Luke to Missy.
"Missy, if you were going to give Luke a ride home, I could do that. No sense you going out tonight. It's getting cold. I can take Luke home, and find out where he lives, in case I need to head to his place one of these days."

Missy shrugged. "It's okay with me, if Luke doesn't mind."

Luke smiled. "I appreciate both of you offering. I hate to bother you for a ride at all. I have a truck, but I never renewed my driver's license. Mason used to do all the driving for the both of us."

Missy nodded, she had already knew about that, but Hank was learning it for the first time.
He smiled at Luke, nodding.
"Come and ride with me. Any time you need a ride you can give me a call also."

The two men said their good-byes to Missy, then got in Hank's SUV. Luke gave him directions.
When Hank arrived at Luke's house he smiled.
"Is this where you live? I love this place. When I was a kid, I use to walk past this place on my way to school. I always thought it was a magic cabin. It looks like it should be sitting up in the woods somewhere. I can remember the woman who lived

here. It seemed she was always out working in her yard. Whenever she saw me, she always waved and said hello."

Luke smiled. "That would have been my grandma Spencer. This house belonged to her, and my grandpa. He passed away back in nineteen eighty eight, and grandma died just three years later."

Hank nodded. "It's nice to keep a place like this in the family. You're lucky. I'm still renting a house. I can't decide whether I want to buy one or not. I should do it, I guess. I'm not getting any younger. I don't have a wife, or kids, so I wouldn't have anyone to leave it to anyway. I own the newspaper office, and don't have anyone to leave it to either."

Luke frowned, then shrugged.
"What about Missy? The two of you seem close, and I know she wants to be a journalist like you."

Hank grunted. "I hope she would be a better journalist than me, but yeah, I've thought about that very thing, more than once. At least I know Missy would continue running the paper, and preserve my legacy."

Nodding, Luke hoped Hank would share that thought with Missy. He sighed.

"Thanks for the ride. I better get inside. Let me know if you learn anything about the Nefasylum building."

Hank nodded. "I will. Good night Luke, it was really good to meet you. I hope we can find out what's happening up at Nefasylum."

Luke smiled. "Same here. Have a good night yourself."

Chapter 5

The next day, Missy was at work. Because Hank had driven Luke home, she wasn't sure if the two of them would be having dinner or not.
After work, she walked down the street.
Seeing Luke seated on the bench he normally used, she smiled, broadly.
"Luke, I'm so glad to see you. The two of us didn't make arrangements for who would be making dinner."

Luke grunted. "You don't have to worry about us having dinner. You must be tried of listening to me talk about Nefasylum, by now."

Shaking her head, Missy grinned.
"Hardly, I want to hear more details. We spoke about a lot of things last night, but I still have about a million questions. Why don't we go to my place? That is, if you don't mind leftovers."

Luke laughed. "If you are still willing to listen to me ramble, then I'm more than happy with having leftovers."

The two walked together to Missy's apartment. Once inside, she got the taco salad she had thrown together the night before, and placed it on the table.

She grabbed a plastic bottle with a light green liquid in it, from her fridge.
She held it up so Luke could see it.
"I made this avocado dressing to put on the salad. I really like it, if you want to try it."

Luke nodded, and added the dressing to his salad. Missy got the two of them cocoa, then sat down across from Luke. She smiled.
"Dig in. When you're ready, I'd like to know more about those experiments up at Nefasylum."
She frowned. "First though, tell me about the building. If there were male, and female patients, did you say you were on different floors?"

Shaking his head, Luke sighed.
"Yes, Neff had us separated. None of the males even saw the females. Except maybe a glimpse here and there. The women, and maybe a few girls, were on the first floor. The men, and yes, a few boys, were on the second. The third floor was set aside for experiments."

Missy frowned. "What about the basement? That's where we saw the light."

A shrug, was followed by Luke staring at Missy.
"I really don't know what was down there. I never was taken down into the basement. I just guessed it was for storage. On the separate floors, each had

one large shower, and bathroom. They were at the end of the long corridor, where the rooms sat opposite from each other."

Again Missy frowned.
"Wait, the rooms didn't have their own bathroom? Doesn't that violate some type of sanitation code?"

Luke grunted. "No bathroom in the patient rooms. We had to get permission to go to the bathroom from an orderly or a guard on duty. Once a week, we were allowed to shower."

Missy stared at Luke, shaking her head.
"Like I said, that had got to be against the law. I think OSHA has laws about sanitation, and things like that."

A grunt sounded. Luke shook his head.
"Nefasylum broke all laws."

Sighing, Missy nodded.
"I read about the people dying in isolation rooms. Were you in one of those?"

Luke nodded. "I was, so were most of the patients. Usually for a twenty four hour stay, sometimes longer. Mason was thrown into isolation for a whole month not long after he arrived at Nefasylum. He was a teenager, and something terrible happened to his mind after that. Before he was put into isolation, Mason was already quiet. Afterward, he didn't speak too much to anyone, except maybe me. I took him under my wing. I couldn't protect him, but I could lend my empathy. I think that helped him get through the four years he was in that place."

Staring at Luke, Missy shook her head, She felt like crying for so many people whose lives were ruined. "Did Doctor Neff use electric shock on patients?"

Luke nodded. "That was a heinous thing to go through. No one's body, or mind is ready to adapt to something like that. When I was taken to the room, I was strapped to a table. I had no idea what was happening. A leather strap was placed between my teeth. Then the electricity was turned on. My whole body tightened up as the flow of power shot through me. I luckily blacked out. I don't know how many times that switch was thrown. For days after, my whole body ached. I know that poor Mason was subject to that ordeal many more times than I was. Like I said before, Neff hated Mason. He had it in for that kid from the moment Mason arrived at the asylum."

Missy sighed. "How awful for Mason. Did you guys ever get a break from these atrocities?"

Luke nodded. "I did. Once a day for an hour or two, I got to go outside. Mason never got that luxury. I was allowed to go out, and sit in the yard. Several of the male patients gathered out there. We talked about some of the things happening, but some things, no one dared talk about. After I was released, I didn't talk about them. The only one I saw was Mason, and he didn't want to relive the horror. He spent a lot of time on his own. He loved to fish. The rest of the time we spent reading, or watching television. We were together the day we heard the Malcolm Neff had been killed."

Staring at Luke, Missy half smiled.
"I bet you were both happy to hear that news."

Luke laughed. "I was, but Mason, he just shook his head. He seemed almost angry. I don't know if he wished he would have been the one to kill the man, or if he actually felt sorry for Neff. He never explained his thoughts about the whole thing, and I never asked."

Missy sighed. "Maybe it's better that way. I still can't believe all of that went on from nineteen sixty five all the way through until two thousand. Those things were outlawed well before the time that place closed."

Luke grunted. "Not to mention, something is still happening up in that place."

Now it was Missy's turn to let out a grunt.
"I just hope that Hank can find a way to get us in to take a look in that building."

Although Luke nodded, and had said he wanted to go up there, he wasn't sure if he actually wanted to see what was in that place, or not.

Chapter 6

Two days later, both Luke and Missy, were given the chance to find out about Nefasylum. Hank had called Luke the night before, but had waited until Missy had come to work, before he had shared with her what he had come up with.
He told her right after she arrived, and was sitting at her desk. He had walked over to her.
"I arranged for you, Luke, and I, to go have a look at Nefasylum. I met with Owen Carter. He is a security Guard, He was hired to do a walk through of Nefasylum a couple days a week. He checks to make sure no one causes trouble up there. The building is owned by an LLC corporation called RavenWing. They have some affiliations with the military, and in turn our government."

Missy frowned. "An LLC, you mean a Limited Liability Corporation?"

Hank nodded. "Yeah, or in other words, a company that makes it hard to trace their involvements. No matter, we at least can walk through that building. I talked with Luke. I also had a talk with George. He is going to keep an eye on the newspaper tomorrow. Then you, Luke, and I, can have the day to go and check out the asylum."

Missy nodded, even though she felt a chill run through her at the prospect.
After work, she walked down to the bench, where she was happy to find Luke sitting.
She sat down beside the man.
"How are you feeling about going up to Nefasylum tomorrow?"

Luke sighed. "I'm not quite sure. I want to know what is happening up there, and I also dread the thought of knowing what is going on, at the same time. I guess either way, it will be good to know. Until then, what should we do for dinner?"

Missy smiled. "Instead of either of us cooking, why don't we go downtown, and grab something to eat?"

Luke nodded. "That's a great idea."

The two walked to a diner not far from the bench where they always met. They only stayed an hour. It was still light enough, that Luke dismissed Missy's argument that she needed to give him a ride, and he instead walked home.
Missy did the same.
Neither of them slept well that night.

In the morning, Hank picked them both up in his SUV. He drove to the Nefasylum building.

None of the three spoke much on the way.

A heavy set man, with dark hair, stood in the front of the building.
Hank parked not far from where he stood.
Getting out of the SUV, Hank waved.
"Owen, glad you're here. This is Luke, and Missy. I told them you agreed to show us the inside of the building."

Owen nodded. "Sure thing. Just don't tell a bunch of people I agreed to do that. I don't know how the people at RavenWing would feel about all of this. Come on, let's get in out of the cold."

The small group waited for Owen to unlock the building, then they stepped inside.
They were surprised to find the interior was much warmer than the outside. Missy was frowning.
"It must cost a lot to keep this building heated. I mean no one is using this place are they?"

Owen shook his head.
"No one is using it right now, but having the pipes freeze up, would cost the company a lot more."

Missy sighed, looking around.
"What about the lights? Why aren't they on as well?"

Owen grinned. "I think that bill is one they wouldn't want to pay. I have flashlights for everyone over at the desk."

The three followed Owen to the desk, which at one time was a reception area. Missy was watching Luke, wondering how he must feel being in this place. Knowing just some of the horrors he had been subject to, she knew it had to be hard.
The bits of light that came in through the dirty windows cast shadows, making the place look creepy.
When Owen handed each a flashlight, they turned them on quickly, grateful for the beams of light that pushed back some of the darkness.
Owen was glancing at the three, but spoke mostly to Hank.
"I have a list of things I need to check out. I normally start up on the third floor, and work my way down. The three of you can head up with me, or you can wander around on your own. Like I said, the power is off, but you should be able to see well enough with those flashlights."

Hanks nodded. "I think we'll just have a look around on our own. How much time do we have?"

Owen sighed, and shrugged. "I'll take my time I can give you a couple hours."
He looked at his watch.

"It's ten thirty. How about we all meet back here by one? Oh, one other thing. This place doesn't have any elevators, so you have to use the stairs."
Owen pointed at a set of closed doors.
"Just follow me through those double doors, then we can go our separate ways."

The trio walked behind Owen. They watched him turn, and head for the wide set of stairs, then Hank, and Missy turned to Luke.
Missy touched his shoulder.
"Are you doing alright?"

Luke nodded. "So far, I'm okay. This is the floor where the females were kept. If you walk up this hallway, you can see the rooms."

Walking toward the hall, Missy pointed at the stairs Owen had gone up.
"These go up, but I don't see any going down to the basement."

Luke sighed. "I'll show you them in a minute. They are further up the hall."

Nodding, Missy moved up the hall, and looked into one of the rooms, then frowned, turning back to stare at Luke.
"Oh my hell, are all the rooms this tiny?"

Luke grunted. "The rooms down here look like luxury suites compared to what I remember from the ones on the second floor. My memory might be a bit off. We'll find out soon enough."
He turned from the room, and pointed toward the end of the hallway.
"That's where the bathroom, and shower room is located at."

Hank, Missy, and Luke walked down, and stepped in. They could see the bathroom stalls were separated, but had no doors.
Missy frowned.
"Not a whole lot of privacy."

Hank was nodding.
"Just a way to degrade people."
He moved to the area where the showers were. The room was large, but it also held no kind of dividers. Moving his flashlight toward the ceiling, Luke could see several shower heads, most were covered with rust spots. Some had broken loose and were dangling down.

Beside him, Missy stared up. She turned to look over at Luke.
"I can't imagine taking a shower, and having others watching. This reminds me of something you would see in a concentration camp."

Luke nodded. "Yes, Nefasylum was a lot like that. Let me show you where the basement stairs are, then I think we should head to the second floor. Sadly, I know much more about that area."

Going to the side of the shower area, Luke headed a short way off, and pointed at a heavy steel, double door.
"That is the way to the basement."

Stepping over, Hanks lifted his flashlight to examine the knobs. He could see the large padlock holding the knobs together. Reaching out he tried to turn one knob, then the other.
"Doors are locked, and then padlocked. Someone doesn't want anyone going into that basement."

Missy sighed. "I hope Owen can get us down there."

Leaving the area, the three used their flashlights so they could see the path back to the stairs.
Missy moved her light quickly back, and forth, her nerves growing more and more strung out. She felt like she had stepped into a horror novel, or scary movie. By the time they got to the top of the stairs, she was certain someone was going to jump out holding a knife. She heard a few noises, and wondered if there were mice, or worse, larger rats, running round. The place wasn't all that clean.

She had noticed the wall paper peeling, and had
seen spider webs in more than one corner.
She looked at Luke, then Hank.
"This building could use a good cleaning."

Hank laughed, shaking his head.
"I doubt if that is a high priority for the owners. I
have to wonder if the basement is in better
condition than the rest of this place?"

Luke frowned. "You couldn't tell looking in from
the outside window. I wonder if Owen will even let
us see it?"

The three had started walking down the hall.
Luke pointed at a room, not quite halfway down.
"That was my room. Mason had the one two down
from mine on the same side of the hall."
Luke moved toward his room, there was no door on
the room. Hank, and Missy walked close behind
him.

Stepping in, Luke moved over to what looked like a
single wide roll away cot. There was no mattress on
the bed, the springs were half broken.
He stepped closer. Luke stood next to the bed, and
pointed at the wall that the long side of the bed sat
against. The other wall was only a few feet away
from Luke's back. He pointed his flashlight at the
peeling wall. He turned to look where Missy and

Hank were standing. The six by eight foot room was small enough that it felt crowded with the presence of the three.
"Look, it's still there. I carved my name there on the wall, not long after I arrived at this God forsaken place."

Luke moved away from the bed, so that first Missy, then Hank, could step forward, and see the name scratched into the wall.
Missy took her phone, and snapped a picture. It was dark in the room, so she used the flash, making sure to capture the name scratched in the wall, in the image.

Hank was frowning.
"How did you manage that? I'm guessing you weren't given any type of metal, or other instruments."

Luke grunted. "You'd be guessing right on that. I had to use plastic forks. I went through several of them, over a month or two, before I finally accomplished the task. Then I had to make sure my pillow was pushed against the name so a guard didn't report me. I can't imagine how Neff would have punished me for that. I'm sure it would have been an isolation period, or maybe an electric shock treatment."
Luke shrugged.

"It would have been worth it. My way of defying the people in charge here. There wasn't a lot of ways to do that."
He sighed.
"Let's go have a look at Mason's room."

Hank and Missy nodded. Missy hurried out of the room first, she felt like the walls were closing in on her. She had never suffered from claustrophobia, but was sure it must feel like what she was dealing with right now. In the hall she drew in several deep breaths.

Luke had already begun walking, but Hank was staring at her. He whispered.
"Are you okay?"

She nodded. "I'll be alright. I can't imagine spending day after day in that room. Did you notice there wasn't any window?"

Hank nodded. "I did see that. One thing I'm sure of, this place was never subject to any kind of health or safety regulations. I can't believe this asylum was allowed to even exist."

Up the hall, Luke had stopped at another doorway. "Hey, are you two coming?"

Nodding, Missy, and Hank moved quickly to where Luke was standing. He moved into the room. Once again, there was no door to the room, and inside, no window was anywhere in sight.
Missy was frowning.
"Did any of the rooms have windows, Luke?"

He shook his head, and let out a harsh grunt. "No, the only rooms with windows were Neff's main office, out by the reception desk, and the room where family members could meet with patients. I don't recall seeing any others. I guess the basement had some, because you and I looked in that one the other night."

Thinking about that, Missy nodded, but frowned. "Strange they would have windows down there."

Luke shrugged. "Maybe Owen will know. We can ask him about that when we see him."
He pointed into the small room, frowning
"No bed in here."

Hank grunted. "The damn thing probably rotted away. I can't believe this whole building hasn't collapsed. Is this room smaller? You'd think without a bed in here, it would look bigger."

Luke nodded. "It's a little smaller. The rooms were a bit different. None of them were large, though.

We had the bed, an extremely flat pillow, and one thin blanket. No sheets at all. We had a chair in here, but it wasn't for sitting on. We had one jumpsuit for wearing, and another that we kept folded on that chair. We weren't given underwear. The jumpsuits were washed twice a month. If you got both sets dirty, you wore them filthy while you waited for laundry days."

Missy stared at Luke, wondering how people could be treated like that. She sighed.
"Did Mason etch his name in the wall like you did?"

Luke frowned. "Not that I know of."
He moved his flashlight around the walls of the room. He shook his head.
"I guess not. It doesn't matter. Mason left his name on his art work, back at my house. That's what matters. Do the two of you want to go look at the bathroom and shower area on this floor?"

Missy got another picture, before she nodded.

Hank nodded. "Let's do that, and then head to the third floor."
He stared at Luke.
"Some day. I'd love to see Mason's artwork."

Luke smiled. "Good, I'd love to show it to you as well."

The three headed to the end of the hall. They found the same type of thing they had found on the first floor. It made Hank, and Missy sad, and angry. For Luke, the memories of the past almost over whelmed him. He had to fight against the horrors, as they drifted into his mind, refusing to allow them to take hold.

Seeing his struggles, Hank grabbed Luke's elbow. "Let's get moving. Owen will be waiting for us in an hour."

Watching the two men, Missy took her time, snapping several more pictures, glad her phone had a good flash on it. The room would have been too dark for getting pictures without that.
She turned, seeing a strange look on Luke's face as Hank grabbed the man.

Luke was frowning, his flashlight aimed at his watch. He was surprised by how much time had passed. Missy must have spent about fifteen minutes taking her pictures. He had been lost in the past and only felt a moment of time sliding by. Shaking thoughts from his head, Luke nodded, and moved to the stairs leading to the third floor.

The three walked up. Luke and Missy together, followed by Hank. The stairway going to the third floor wasn't as wide as the other they had come on. It was as dark as the rest of the building, and flashlights were kept trained in front of the three as they made their way up.

At the top, Luke headed straight to the far end of the hall. The room he went to had a large, heavy door. Luke pushed it open, then stood back, and drew in a breath.
"This is the isolation room."

Missy's blue eyes grew wide. She wasn't sure she even wanted to look in the room. She pushed back her long, brown hair, and let out a long sigh. Moving to the corner of the room, she turned her flashlight so the beam focused on a dirty, torn mattress on the floor. She was sure the darker spots on the material were blood. She looked at the wall just above the thin mattress, and grunted.
"Oh my hell, are those chains?"

Luke nodded. "I don't think Neff wanted us walking around while we were in isolation."

Missy grunted again.
"Would the two of you point your flashlights in this area? I want to get a good picture of this spot."

The men nodded, and lifted their lights. When they did, a large rat, ran from the corner of the mattress, Missy let out a scream. She dropped her phone, covering her mouth with both hands trying to stifle the noise. She was afraid someone, living or dead, would hear her, and come running.
A moment later, she felt calm enough to drop her hand, and feel around for her phone. Grabbing it, she took several pictures, then turned to the men. "Sorry about that, I guess I'm more nervous than I though. Let's get out of this damn room."

Both men nodded, glad for the suggestion. The three headed to the hallway, and stood silently a few moments. Finally Luke sighed.
"There are a couple more rooms to look at, if you still want to see them."

Missy nodded. "I think we should. I feel better now."

Hank also nodded. "We might as well have the whole tour."

Luke shrugged, and let out a deep breath.
"Let's do it then."
He moved down the hall, with Missy and Hank staying close to his side.

Missy noticed the rooms on the third floor had doors on them, unlike the ones on the floors below. Besides the isolation room, there were only two other doors. Luke went to one of them. It was a large double door. Luke pushed it open.
"This is where we were brought for the electric shock treatment."

The trio stepped into the room. Only one table sat in the room. It was a metal table on wheels. Three light beams played on the table. The leather straps still hung on the edges. Two on the top, and two on the bottom. If someone was on the table, the straps would have sat where a person's arms and legs would have been. It was easy to imagine someone being tied down.

Missy moved closer. She could see the straps were now cracked and split in places. She reached out a shaking hand to touch one, then pulled it back, before making contact. She knew if she touched the damn thing, she would vomit. Instead, she pulled out her cell phone, and turned to Luke and Hank. "Shine your lights on this."

As soon as they did, she took her pictures, then stepped back. She looked at Luke.

"Is there just one more room?"

Luke nodded. "Yep, the operating room. Come on."

Back in the hallway, the three moved to the last room. It also had large double doors.
Luke started to push them open. Seeing he was having a hard time, Hank stepped up and helped the older man.
Both men, leaned their shoulders into the doors.

A moment later they creaked open. The sliding metal grinding against the peeling linoleum floor. Standing back a few feet, Missy took in a breath, then held it. She didn't know what she expected to find in the room. Maybe a bloody corpse, or even a doctor, standing in the room, holding a beating heart. His maniacal laugh, filling the room.
Instead, the deathly quiet, was somehow, as bad.
Missy frowned, trying to peer into the darkness.
She lifted her flashlight, and waved it around.
The room was huge. Much larger than any room they had been in so far. The ceiling was higher. Large lights hung down. Now devoid of power, but once they must have been brilliant. Missy could almost see Doctor Neff standing over a patient, doing some awful experiment. Maybe even a castration. She felt her stomach lurch, and covered her mouth. She swallowed hard a couple times. Then she drew in deep breaths, before stepping forward to join Hank, and Luke.
The two were talking, oblivious of the problems Missy had been having. She was thankful for that.

Hank was nodding, at what Luke was saying.
"The sterilizations, castrations, lobotomies, and other experiments happened here. I'm sure many of them, I didn't even know about. I heard rumors here and there, but I didn't talk with many of the other patients. Mostly, I just spent time with Mason. But hell, he had enough experiments done on him over the years to fill several books."

Shaking his head, Hank scowled.
"This is just reprehensible. If they are doing something like this down in the basement to this day, they have to be stopped."

Luke nodded. "The only way to know that for sure, is to go down there."

Hank nodded.

Missy sighed. "Before we go, let me get some pictures of this room."

The two men used their flashlights to help Missy, then the three made their way back down to the first floor to meet up with Owen.

While they waited, Missy also got pictures of the reception area, and the room where family met with patients, to show the differences between the main floor, and the rest of the asylum.

Looking at her, Hank was grinning.
"I'm glad you thought to get pictures of everything. I wish we had a better record of things back when Luke and the others were in this damn place."

Before Missy could answer, Owen stepped into the room, and nodded.
"Glad the three of you are back. I think it's time for all of us to get out of here. I don't want to hang around this place any longer."

Hank frowned. "I thought you would take us down into the basement."

Shaking his head, Owen frowned.
"I never go down there. Looking after the basement isn't part of the security job. In fact, the basement is padlocked. I don't think the regular door locks were working well enough, but someone added the padlock before I ever began working here. Nothing down in that place to see anyway."

The three stared at Owen, but none of them challenged what he was saying. They also didn't mention they had examined the doors, and the lock on them. They knew any of them questioning Owen's explanation, would be an exercise in futility.

Instead, Hank just nodded.
"Well, we appreciate you letting us roam around in the building. I know you could get in trouble for doing that."

Owen nodded. "Damn right, I could."
He moved to the front door, and escorted the three outside.

As soon as they were far enough away that Owen couldn't hear what they were saying, Hank turned to Missy and Luke.
"Why don't I treat the two of you to a late lunch, or an early dinner."
He looked at his watch.
"It's time for either. I think the three of us have a lot to discuss."

Nodding their agreement with what Hank said, the trio decided on a restaurant, then began walking.

Chapter 7

Stepping into the restaurant, Hank asked for a booth. He wanted the three of them to be able to talk with some privacy.
Sitting down, they ordered the meal, and some drinks. They didn't speak until the waitress brought back their drinks, then left them, to wait for their food.

When she left, Missy was the first to break the silence. She was staring at Luke.
"I don't know how that place felt to you. It scared me to death. It must have been a lot worse for you. Dredging up those old memories."

Luke shrugged. "It was bad. A lot of things from the past came rushing back. The worst were the memories of Mason. Not just the things he went through, but the times he and I spent together. Not just up in Nefasylum, but after we got out of the hideous place. The two of us at least had each other to lean on. I can't imagine how others dealt with the things Neff did to them. Physically, and emotionally, that is a heavy weight to carry."

Missy nodded. "I wonder how many people who were in that place are still alive?"

Hank sighed. "I can't imagine many would want to talk about what they went through. If they would be willing to, we might have a chance of stopping whatever is happening. To stop the things going on, up in the basement now, from continuing."

Looking at both men, Missy's eyes narrowed.
Her mind was whirling with ideas.
"First, we need to get in the basement, and see exactly what is going on."

Luke's light blue eyes opened wide.
He shook his head.
"I don't think that's a good idea. Especially for you. It was one thing for me to take you along to take a peek inside that place, it's a whole different story to think about trying to get inside. Those kinds of people aren't playing games. They don't care who might get hurt or even killed. I told you about some of the things they did. I can never put into words the terror, the anguish, the horrible humiliations, we suffered."

Missy nodded. "I believe that, and that is why we can't allow those people to even think about trying to do those things again."

Before Luke could reply, the waitress walked toward them carrying the food they had ordered.

The talking stopped, as she placed the food on the table. Missy was thankful for the interruption, but she also knew that wasn't going to stop Luke from trying to prevent her from trying to get into the basement of Nefasylum.

Instead of talking while they ate, the three focused on their food. When they finished eating, they headed back to where Hank had parked his SUV. He looked at Missy, and Luke.
"Would the two of you have time to come to my place? I've been checking into the people who bought the Nefasylum building. I also might have some information on a few of the people who spent time up there. I was going to wait before I shared it, but after Owen spoke the way he did about not going into the basement, I think it's time we all talk some more about what's going on in that place."
Hank turned to stare at Luke.
"I know you're worried about Missy. You should know she is someone I would choose to have my back, no matter who I'm fighting with."

Letting out a sigh, Luke nodded.
"Okay then, let's go to your place. I'll try and have an open mind. Just remember. I'm still a little old school in my thinking, and in my education. I still want to be the protector of women, and I can hardly turn on a computer."

Both Missy, and Hank laughed. Reaching out, Missy took Luke's hand.
"You can still be my protector when I need it, and I can be yours, when you do. As for the computers, I'm still learning about them myself."

Hank was still laughing.
"Okay you two, hop in. Let's get to my place, and see if we can figure out what the three of us can do."

Driving to Hank's house no one spoke much. When they arrived at the house, Luke pointed at the rather plain, white home.
"Is this your place?"

Hank modded. "Yeah, I like your place better. I've been thinking about buying a place, but never can decide if I want to or not. I know I want to live here in Mansfield. I love having the newspaper. I mean, I couldn't imagine doing anything else. If your place ever comes up for sale, let me know I'd buy it in a heartbeat. This house just isn't my home."

Luke laughed. "Not until I leave this earth. But before that happens, I'll make sure you have the first choice to purchase the place, for sure."

Listening to the two men, Missy was shaking her head.

"I don't even want to hear talk like that. Remember Luke, you're sixty eight, not eighty six."

Luke laughed, remembering his words to Missy. "I do remember saying something like that. And you're right, we shouldn't be talking about things like that. We have enough dreadful things on our plate right now."

Hank nodded. "Let's go in, and we can discuss those things."

Following Hank, everyone headed inside.
He took them into what could be called a den, or a library. Papers were strewn across the top of a large table. Hank pointed at them.
"Ignore the mess, I have a system that no one can figure out. Find a chair, I'll go put on a pot of coffee."

After Hank left the room, Missy and Luke found chairs. The two looked at each other, then gazed around the room.
Missy's eyes narrowed, staring at the mostly bare walls.
"When Hank said this place wasn't his home, he meant it. If you didn't see that table covered with papers, you'd hardly even know someone lives here."

Shrugging, Luke nodded.
"I can understand that. I'm lucky to have my house. It always felt like home to me. Even when I didn't live there. I was lucky to have my grandma Spencer. She was always my biggest supporter, and defender. She believed in me, even when I didn't believe in myself."

Missy smiled. "She must have been a wonderful lady."

Nodding, Luke smiled, thinking about the woman. Before he could reply, Hank stepped in.
"Come out into the kitchen, and grabbed a cup of coffee. We can bring the drinks back here, and then go over some of the information on Nefasylum, and things related to the place."

Nodding, Missy, and Luke followed Hank to the kitchen. The room they stepped into didn't look any more lived in, than the den, they had been sitting in. With their cups filled, they returned and sat back down. Hank had Luke and Missy pull their chairs closer to the table.
He moved a few papers to reveal a laptop beneath. He shrugged at their surprised looks.
"I knew it was there, believe me."

Missy laughed. "I've seen your office. I know how you do things. So, what have you found. I know you said that this company RavenWing is an LLC, and has some dealings with the military, which means the government, as well."

Hank nodded. "RavenWing did some work at a military base over in Wyoming, the east side of this state, and also in Utah, where they hid some interesting things. The military was doing work into what many believed was a connection with UFO's or whatever they are calling them. It was reverse engineering, and experimenting with devices that weren't from our world. I don't know how true that was. There were also stories that RavenWing got a hold of some of the papers that were found after Nikola Tesla died, that had to do with free energy and even time travel."

Luke stared at Hank.
"Now that would be interesting reading. I was always fascinated by Tesla."

Hank nodded. "I think the government is hiding too many secrets. I also think they use the military to ensure those secrets stay hidden."
He shrugged.
"Anyway, about three years ago, the RavenWing Company bought that building. It was so dilapidated, many thought it would be torn down, and something new put in place. The company just left the old building in place. No one saw anything

happening up there. Most people didn't even think about it. Not until the two of you talked. Now, everything has changed."

Missy nodded. "Damn right. So, now what?"

Hank smiled. "Actually, I've been giving that a lot of thought."

Both Missy, and Luke nodded. Neither spoke, as they waited for Hank to share his thoughts.

Smiling at the anxious looks, Hank drew in a breath, and began talking.
"What I was thinking, was that we should write a partial story about what happened up at that place in that past. I don't want any mention of what we think could be happening up there now. I'd like to ask for the public's help, See if anyone has stories they would be willing to share. Any experiences, either their own, or some they have heard from others. Whether something happened to family members, friends, or even people who might have worked in that place."

Luke sighed, and nodded.
"You might hear from patients. After all this time, they could feel it's safe to talk. As for people that worked there, yes, there were some, who didn't like what they saw. I met a few who were disgusted by

what was going on. They were told not to speak about what was happening in Nefasylum. Even those who quit were threatened if they talked. I have a feeling, even after that place closed it's doors, people still lived in fear of the repercussions for talking about the atrocities. If those people see your paper is willing to print what happened, they might feel now it is safe to talk. They'll know the government's long reach is maybe not gone but has lessened."

Missy nodded. "That means keeping quiet about our theory, that something could be happening up in Nefasylum, is all the more important."

Luke was frowning. "I don't like the connections the new owners have with the military. That makes all this really dangerous."

Missy grinned. "It just means we just have to be more careful trying to get into that basement to get a look around."

Both men turned to Missy, staring at her like she had lost her mind.
Luke was frowning.
"I don't think going up there is a great idea. Not for a while anyway. I think we should let Hank write up this article in the paper he's talking about first. That may stop anyone up at Nefasylum from doing

anything right now. They might be a little nervous seeing what Hank is doing."

Missy sighed. "Okay, but I still want to go up there. I'm afraid they will start up those experiments again. Luke, you better than anyone, knows what that means. I'm sure you don't want anyone else to go through what you, and Mason did."

Luke nodded. "God forbid that ever happens to anyone. I don't think I could live with myself if I could prevent something like that happening and didn't do all I could to stop it."

Reaching out, Missy placed her hand on Luke's. "I know you aren't that kind of person, Luke."

Luke shrugged. "Sometimes I wonder. I had such a hard time dealing with everything I went through. Then trying to help Mason with what he had to deal with. The two of us were overwhelmed with just getting through the day we didn't have time for much else."

Staring at Luke, Hank was shaking his head tiredly. "No one should have that kind of thing to deal with. The people who did those things should have been behind bars. Nothing like that should ever be allowed to happen again. I hope we can prevent those atrocities."

Missy was nodding, then she sighed.
"It's been a long day. I wonder if I could get a ride home? I think I could use some rest."

Luke nodded. "That sounds like a great idea."

Hank grinned. "Sure thing. I can give you both a ride."

He turned to look at Missy.
"Would you consider helping me write up that article? I think a woman's touch would help a lot. You saw that place. Even empty it invokes some strange images in someone's mind."

Missy smiled. "Do you really mean it? I would love to do that. You're right about those images. I wonder if I will even be able to get the rest my body is crying out for."

Hank nodded. "Of course I mean it. You should go home, take a hot bath, and watch a comedy or two on TV, maybe you can rest then,"
He turned to Luke.
"I might give you a call as well. You know more than I would about what it was like in there. I won't use your name in any articles I write. I do feel I could get a more accurate story of how being it that

institute felt. I mean if you don't mind. All of this must be hard on you."

Luke sighed. "If you think it will help, I don't mind at all. Especially if whoever is in that place is thinking of doing it all again."

Hank nodded. "Great, let's get the two of you to your homes then."

Chapter 8

At the newspaper office, Hank and Missy were in Hank's office. They had asked George Benson to keep an eye on the front office while they worked on the article about Nefasylum. They hadn't told him what they were working on, just that it was a special project the two had decided to work on. Hank had told George that he would give the man his own project in a few weeks for his helping out. George, who was planning on retiring soon, didn't really care about having his own article, but still the thought of doing that, intrigued him.

Hank had already made several calls to Luke. He wanted Luke to come in to the office, then decided he didn't want Luke to be seen at the newspaper office. He didn't want anyone associating Luke's face with the article. The possibility was remote but still, even if unlikely, it could put Luke in danger. The man had been through more than his share of heartache, and horror.

In the days since Missy had found Luke sitting on the bench with tears in his eyes, Luke had cremated his friend, Mason without telling Missy, or Hank.

The paper didn't even run any type of obituary or death notice.
When Luke finally mentioned it, Missy and Hank, assumed dealing with Mason's burial was something personal that Luke hadn't felt like sharing with them until it had been finished.
Both of them wished they had been there though, just for an added comfort for Luke.
Mason had been cremated, but not buried. His ashes were in a container out in Luke's garage. The remains took a special place on a shelf below one of Mason's pieces of artwork.
Luke had told the two about that, when he had been talking to Hank, on speaker phone, when Hank had called to get information about Nefasylum.
Missy had cried.

At the newspaper office, it took Hank and Missy two days of pouring over whatever information they could find, before they felt the article they had put together was ready for print.
Before they made the final decision, Hank called Luke.
When Luke answered his phone, Hank drew in a deep breath, before speaking.
"If you have a few minutes, Missy and I wanted you to hear this article, before we send it to print. If you give your blessing, then it will definitely go to print."

In his kitchen, Luke pulled out his chair, and sat down, preparing himself.
"I have time. Go ahead."

Hank cleared his throat, and read the story. By the time he was done, Luke took a minute to get his emotions together. He sighed, and nodded.
"I think you captured that place well. If you are asking if I approve, then I say, hell yes, go ahead and print the article. You, and Missy did a great job."

Both Hank, and Missy, were pleased with Luke's praise. Missy called out, so Luke could hear her.
"Thanks Luke. I hope I'll see you after work waiting on the bench."

Luke laughed. "I'll be there."
He hung up the phone. Then stayed seated at the table. Allowing all the emotions he had pushed back, while he talked on the phone, finally settle around him.

At the office, Missy turned to Hank.
"Looks like we go get set up to print the next edition of the paper."

Hank nodded. "Yep, and then we let George take a few days off. He deserves it. He probably wonders if the two of us will ever show our faces outside this room."

Laughing, Missy stood, and walked from the office.
She looked over at George, and smiled.
"How are things going?"

George shrugged. "Not too bad. If you stepping out of that room mean you, and Hank are done with you're article, then I'd say things are looking up."
Nodding, Missy laughed.
"We are finished. Just have to set things up for printing."

Hank also stepped into the main room.
"Missy's right. Thanks for all the work you've been doing George. Why don't you go take your lunch break. If you want any other time off, just let me know."

George nodded. "Just lunch for now. Thanks Hank."

Missy, and Hank set things up for the next day's edition, while George was gone. They had him look at it when he returned.
George read the article, a sober look on his face.
When he was done, he turned to look from one face to another.

"That place never should have been allowed to begin operating. It's a savage, dark spot on the history of this area. To think about the time when that happened is even worse, Most places had stopped those types of practices. I never understood why there wasn't some kind of regulation on that asylum."

Hank nodded. "You're right. Everything that happened there is an offense to normalcy. No one should be treated like that, and no one should have allowed it to happen. Hopefully something like that will never occur again."

After work, George headed out first. Instead of going to his SUV, Hank agreed to the offer to have dinner with Missy and Luke.
The two walked from the newspaper office together. They walked to the bench where Luke was sitting.

He smiled seeing the two of them.
"Well, isn't this nice. Hi Missy, hi Hank. What's going on?"

Hank shrugged. "Missy asked me to join the two of you for dinner. Hope you don't mind."

A chuckle sounded. "Sounds great to me. More work for Missy though. She's the one who would

mind. If she asked you, then it must be okay."
Luke stood. "I really think the two of you did a great job with that article you'll be putting in the paper."

Hank sighed. "I hope so. It comes out first thing in the morning."
The three began walking. Hank continued talking. "What I really hope is that people reading it will come forward with more stories."

Missy nodded. "At the end of the article, we ask for people to do that. We also added that people can remain anonymous. All of us know that people will still be afraid of consequences for sharing what they know about that place, and the repulsive things that went on."

A sigh was followed by Luke shaking his head. "And that might still be happening. I hope we are wrong in our thinking about that."

The three had reached Missy's apartment. She walked slightly ahead and unlocked her door. Holding it open, she stepped back, and motioned for the two men to enter ahead of her.
"Head out into the kitchen. I'll grab us something to drink. Then I'll throw us another of Mrs. Forsyth's casseroles in the oven. I swear she must cook hundreds of them at a time."

Luke laughed. "I'm just glad she's a good cook."

Missy nodded. "She is that. I've never gotten a meal from her that wasn't delicious."

A half hour later, Missy pulled the meal from her oven, and the trio sat down to eat.
While they ate, Missy was staring at Luke.
"I'm sorry I didn't meet you sooner. I mean I know the two of us passed the time of day, but I wish I would have talked more extensively with you before Mason passed away. I feel so awful that I never met him. I'm also sorry I wasn't by your side when you had to handle his cremation. I feel terrible that you did that alone."

Luke shook his head.
"No need to feel bad about that. Mason and I spent most of our lives with just him and I being together. His dying, and me putting him to rest, was something I needed to take care of on my own. If I would have needed help, I would have called you. I mean that. I feel close enough to you, and to you as well, Hank, that I can call both of you my friends. I wanted to do that on my own. Some things, a person feels better doing by themselves."
He shrugged.
"I don't know if you can understand what I'm trying to say."

Nodding, Missy gave Luke a sad smile.
"I think I do understand. I'm glad you told me. Let's push all that back, and go on to other things. Have you thought any more about going and taking a look at the basement up at Nefasylum?"

Both Luke, and Mason, dropped a hand down on Missy's table. The slamming sound, made her jump. Luke groaned.
"We need to stay away from there, for now."

Hank nodded. "The article coming out is going to have those people watching for anybody looking around."

Missy nodded. "That's why I was thinking we could go up there tonight. That article won't be published until morning. We could sneak up there before it is published. That would be our window of opportunity."

Luke grunted. "And here I was thinking that pretty little head of yours had a good brain inside. How did I get that so wrong?"

Hank nodded. "You know what? I thought that same thing myself. All this time, she had me fooled too."

Missy rolled her eyes.
"Just for that, maybe I should let the two of you go without dessert."

Then she began laughing.

Chapter 9

A week later, as decorations for Halloween were just beginning to appear, the first call came into the newspaper office.
Missy was at her desk, and took the call from a woman who sounded nervous.
"My name is Tara Masterson. I'm calling about the article you ran in the paper. The one on that asylum up on the hill."
The woman groaned.
"Well not me exactly. It was actually my husband's grandmother who was up in the horrid place. Could Kyle and I come and talk with someone?"

Missy nodded, as she wrote down the names on a pad in front of her.
"Yes, is there a time that's good for the two of you?"

Tara sighed. "Could we come in today? Kyle gets off work at three. We could come in right after that."

Missy was grinning.
"Sure, any time is good."
She frowned.
"Would Kyle's grandma be willing to come in as well?"

Tara's voice broke.
"No, His grandma passed away."

Missy sighed. "I'm sorry to hear that. Could the two of you be here at three thirty?"

Tara nodded. "We'll be there."

Hanging up the phone, Missy hurried over to Hank's office. She knocked on the side of the open door, then stepped in as Hank looked up.
A smile on her face.
"Don't make any plans for this afternoon."
Missy's smile got broader as Hank frowned.
"We got out first call on that story."

Staring at Missy, Hank lifted an eyebrow, with curiosity.
"Did they give you any information?"

Missy shook her head.
"Not really. A woman named Tara, said her husband Kyle's grandma, was up in Nefasylum."

Hank grunted. "Now that sounds intriguing. What time are we looking at?"

Missy smiled. "Three thirty."

Now, Hank was also smiling.

For Hank and Missy, three thirty took a long time to arrive. For Tara, it seemed to come too soon. For Kyle, the moment was both too long, and too quick. Since he was a child, he had heard the tragic tales of what his grandmother had gone through. The things she had suffered at Nefasylum. The only good thing that had come from the experience was his mother. Now, he would be sharing the whole tale with strangers, maybe many others, if he allowed the newspaper to print what had happened. He and Tara had talked about it with his mom, Leigh. The three of them had decided, the story should be told. Mainly so something like it would never be allowed to happen again.
Standing on the sidewalk, outside the Mansfield Blotter, both Kyle, and Tara Masterson stared at the printing on the window, before Kyle sighed, and pulled the door open.

The two stepped inside. They looked at the woman who stood up from behind a desk, she smiled.
"You must be Tara, and Kyle. I'm Missy. I talked with Tara earlier."
When Tara nodded, Missy stepped over and shook both the couple's hands.

"If you follow me, I'll take you to meet my boss, Hank. We can talk in his office. The article you read was put together by the two of us."

Following Missy, the two went in to Hanks' office where introductions were made, before the two were given a chair, and then offered drinks. Both asked for bottles of water. Missy hurried out and grabbed them, then returned to the room, closing the door behind her.

After Missy sat down, Hank cleared his throat.
"I want the two of you to know, nothing you tell us will go farther than this room. I also want you to know we have talked with others who have spent time up in that house of horrors."
Hank sighed, not sharing the fact, it had only been Luke they had spoken with.
"What happened there broke all rules against humanity. The people who were put into Nefasylum were subject to appalling things no one should have gone through. Missy said that Kyle's grandmother was in that place, and has since passed away. I'm so sorry that she was ever there, and sorry for the loss you both have suffered."

Both Kyle and Tara were nodding.
Missy looked at Tara.
"I talked to you, so I don't know who would like to tell us about what happened. Hank and I want both

of you to know, that we are open to hearing what you have to tell us. We also want yo assure you nothing you say will be beyond belief to us."

Clearing his throat. Kyle sighed.
"I should be the one to talk to the two of you. I have to admit, I didn't want to do this. It was Tara, and my mom, who convinced me I should. My mom, her name is Leigh. She said, my grandma was owed this. My grandma, Rose Marie Farrell, she was only sixteen when she ended up in Nefasylum. She didn't do anything wrong."
Kyle grunted, shaking his head.
"Not really. She skipped a few classes at school. Hell, who hasn't done that? The counselor at the high school thought she should be evaluated up at Nefasylum. I think that place had just opened. That was in nineteen seventy six. Grandma was sent up to Nefasylum. Shortly after she arrived, she was raped. Not once, but repeatedly. Grandma said both the doctor up there, and some orderlies raped her. The doctor's name was Neff. I wish I would have been around at the time."

Beside Kyle, Tara patted his leg, and shook her head.
"I'm glad you weren't. You'd be sitting in a jail cell for murder."

Kyle blew out his breath. Giving Tara a small grin, and nodding.
He turned back to face Missy and Hank.
"After grandma had been in that place less than a month, she knew she was pregnant. She was kept in that place until her baby was born. My mother came into this world in March of nineteen seventy seven. My grandmother was thankfully released from that house of horrors then. She was told if she talked about what happened, the baby she had come attached to, would be killed, and my grandmother's parents would also be harmed. My grandmother had no doubt that the people from Nefasylum would do those things. Those people were deplorable, and not above murder. The whole story didn't come out until after I had graduated from high school."

Beside Kyle, Tara was nodding.
"Poor Rose Marie. I think until the day she died, she took some of the blame for what happened. She thought her skipping school, all those years ago was the catalyst that started the horrific chain reaction."
Tara shook her head.
"Can you imagine?"

Missy stared wide eyed at Tara.
"Oh my gosh. How awful is that? That poor woman. I hope Doctor Neff and those others who did similar things, and those who allowed them to happen, rot in hell."

Nodding, Tara was silent a moment, then she began to sing, quietly. She had a sweet voice.
"They created wounds, and broke you down, but not your loving spirit, or your caring heart and soul. Tears were shed for the hurt they caused and the things they stole. Now it's time to spread you wings, sweet Rose Marie. Now your spirit can truly fly. Soar through the sky. Heaven is open to thee."
Staring at Tara, Missy saw a tear slide down the young woman's cheek.
Wiping the tear away, Tara grinned slightly.
"That's the chorus to a song my friend wrote. She sang it at Rose Marie's funeral. Everyone who met Rose Marie loved her. She was just so wonderful." Tara's voice broke up.

Kyle slipped an arm around his wife's shoulder. He nodded at what Tara was saying.
"Tara's right. My grandmother was a wonderful woman. She had such a big heart. Despite what she went though, she loved everyone she met. I hope you will decide to put her story in your paper. I also hope others will feel they can come forward and tell there own. Hopefully that will help things like what happened up in that place from happening ever again."

Behind the desk, Hank was nodding. He had waited a moment to speak. Watching Tara with that tear slipping down her cheek had almost made him cry. "Rose Marie's story will be put in the paper as soon as we can get it written up. I just would like to ask if you want us to keep her name out of the article. We can write the story with no name, a made up name, or with Rose Marie's name. Whatever you feel the most comfortable with."

Kyle frowned. "I don't want to use her name. Not right now. We didn't bury her all that long ago. One day I might have her name put out there for others to know about, but not just yet. Can you just tell the story anonymously? I'd rather not use a phony name either."

Hank nodded. "Yes. We can do that. I appreciate you both telling us your grandma's story. I know how hard this must be."

Kyle stood, then Tara. Reaching out, Kyle shook Hank's hand. He sighed.
"Grandma did the hard part, not us. Thanks for printing the stories about that atrocious place, and the repugnant history of it. Those things are long overdue."

Also shaking Hank's hand, then Missy's, Tara also nodded.
"Yes, it's a good thing you both are doing."

When the two walked out of the office, then the building, Missy turned to look at Hank.
"Wow, I wasn't expecting that. Can you imagine what that family has gone through? I mean we're talking generations of pain. Oh hell, I feel like crying for all the families who had people up in that place. Oh hell."

Hank nodded. "Exactly."
He shook his head. Then he sighed.
"Looks like I better go let George know he'll have to do a few more long shifts, and some covering for you at the front desk."
Hank looked at his watch.
"We can get started on this story. At last get an outline put together before we leave for the day. Are you making dinner for Luke?"

Smiling, Missy shook her head.
"No, Luke offered to make the dinner for the next couple of nights."
Then she laughed.
"I'm thankful for that, especially now. Why don't the two of us come to work a couple hours early? That way we can get a big chunk of this article finished before George shows up for work."

Hank nodded. "Great idea. I'll go talk to George, then we can head back in here to work on this new article."
He sighed. "I wonder how many more people will call, or come in with stories like Rose Marie's?"

As Hank left to go talk to George, Missy thought about the question he had just posed, wondering if Hank was going to need to hire extra help. All because of Doctor Neff's diabolical dark works. Dirty activities that might still be happening.
A few hours later, Missy left work. She had asked Hank if he thought it would be okay if she talked with Luke about what the two of them had learned about Rose Marie Farrell.
Hank had told her, yes, as long as she didn't mention names.

After going to her apartment, Missy headed to Luke's place. When she knocked on the door, he answered quickly.

Taking one look at her face he frowned.
"Are you okay? Get into the kitchen, and sit down. I'll make you a hot drink. Maybe with a small bit of alcohol in it. Not much, because you're driving. A hot toddy will fix you up."

Going in the kitchen, Missy sat down. It was just a few minutes later, that Luke handed her a large mug. Missy smelled it and smiled.
"What is this?"

Luke laughed. "My own recipe. I use coffee, a bit of rum, a touch of chocolate, and a dash of sweet cream. There is one more ingredient, but that is secret. Try it."

Taking a drink. Missy smiled.
"I love it. Thanks."

Luke laughed. "Your welcome. Now, tell me what's going on. You look pale, and to tell you the truth, a bit like your best friend died. The food needs to cook another fifteen minutes, so we have time to talk."

Sighing, Missy shrugged.
"The newspaper article brought in the first personal inquiry from someone with a story to tell. Actually the story is about a loved one that passed away."
She stared at Luke.
"Her family asked that we not mention names for now. I will say the woman was only sixteen when she was shipped to that awful place. All she did was skip a couple of classes at school. Can you believe that? The counselor who had that bright idea, should have been sent up there, instead of a young

girl."

Missy shook her head, drawing in a breath, trying to quell her anger.

"That girl was raped, by the doctor, and by others. She ended up pregnant. Luckily, after the baby was born she was released. Of course she was threatened if she talked about what had happened. Even her families lives were threatened, and the life of that innocent child. The man, who told Hank and I the story, was the grandson of young woman who had been raped. She has since passed away. That horrid doctor ruined the lives of three generations. Mentally, and physically, he destroyed the hopes and dreams of so many people."

Luke was nodding.

"It wasn't just women who were sexually abused in that place. Even if the patients weren't threatened, a lot of them wouldn't have spoken about what happened. So many victims of sexual abuse feel like they are to blame. They are shamed into feeling that way. That makes the abuse all the worse. Those things are happening every day, and as we speak, they continue."

Nodding, Missy jumped in her chair when the alarm on the stove beeped. Her nerves were balled up so tight, she almost screamed.

Standing, Luke placed a hand on her shoulder.

"Let's have some dinner. I made a roast, mashed potatoes and gravy. What my grandma would call comfort food. I think we could both use a little of that right now. Hell, I even made apple pie for dessert."

Nodding, Missy laughed.
"Apple pie is the best extra comfort ever. That sounds great to me."

Going to the stove, Luke pulled the roast from the oven. Before he dished up the meal, he turned on the radio, and turned the dial to his favorite station that played the mellow sixties, and seventies songs. He hoped the slow music would help further soothe the senses of both of them. He wondered in the coming days how much of that would be needed when more people began coming forward with the horror stories of the asylum that sat on the hill overlooking the town of Mansfield.

Luke and Missy ate dinner, then sat in the kitchen drinking cocoa, this time with no sort of alcohol added. The two avoided any talk of Nefasylum, for now. They instead enjoyed the soothing music, until Missy yawned, and decided to head home and try to get some sleep.

The next morning when Missy got to work, Hank was standing at the front door of the newspaper office waiting for her.
Missy frowned looking at her watch.
"What's going on? I'm not late. In fact, I came early like we agreed on yesterday."

Hank nodded. "Good thing too. We have three people that will be here at nine sharp. They want to tell us their story."

Blue eyes widened.
"Oh hell, really? Who are they?"

Sighing, Hank shrugged.
"Grab yourself a cup of coffee, and come on in my office. I'll explain what I can while we wait for them. Our working early didn't go as I thought it would."

Sighing, Missy got her coffee ready, then headed to Hank's office, and took a seat.
She sat on the edge of it, leaning toward her boss.
"Okay, I'm all ears."

Hank nodded. "I got a call last night at home. The phone here is linked to my home phone, just in case someone calls after hours. Anyway, when I answered a mans' voice came on the line. He said he was Corporal Langdon. The Corporal said he, and two people in his group, had a story to share with the paper. He asked if their names could be kept out of any article that was written. When I told him yes, he asked if the three of them could come in at nine this morning. He didn't give me any other information. I wish I had more to tell you about all of this."

Missy sighed. "I guess we'll both find out, soon enough."
She pointed through the small window in the door. "Three people just came in the newspaper office. If I had to guess, I'd say they have arrived."

Standing, Hank stepped out. He looked at the three. A frown on his face.
"Corporal Langdon?"
One of the men nodded. He had sandy hair and brown eyes. He stepped forward and shook Hank's hand.
"Yes sir, I'm Corporal Langdon, or was. Since this story happened, when we were all in the military, the three of us thought we should tell you our names, and the ranks we had, at that time."
He turned and pointed at the woman, who had brown hair, and eyes.
"This is Private First Class Scott."

April Scott stepped forward to shake Hank's hand.
"Good morning sir."

When she stepped back, Rod Langdon pointed at the other man.
"This is Private First Class Benson."

With his hand out, Daniel Benson stepped forward. His blonde hair was in a crew cut. The blue eyes narrowed slightly.
"Good to meet you sir."

Taking the man's hand, Hank nodded.
"Would any of you like some coffee, or something else to drink?"

When the trio all shook their heads, no, Hank nodded his.
"Okay then, let's go in my office. My assistant Missy is waiting for us."

Stepping in, Hank introduced the three. Shaking their hands, Missy asked if she could call the three by their first names. She grinned.
"I hope you don't mind my asking. That just seems easier to me."

Agreeing, the three gave their names. Missy smiled.
"Thanks. Since you are in here together, I assume what happened was when you were in your capacity in a military group."

Rod nodded. "Yes, there were seven of us in the squadron, and also the lieutenant colonel. Only the three of us are willing to talk about anything that happened. In fact, I don't think revealing we were in that squadron is a good thing. Can you write this story without releasing that information?"

With a slight grin, Hank nodded.
"I think we can work our way around that."

Rod sighed. Both Hank and Missy could see the relief the man was feeling at Hank's statement.
He began talking again.
"Alright, I guess I should start from when the Lieutenant called the squad into his office. This was about twenty eight years ago."

Staring at the man, and the others with him, Missy guessed the trio must be around Hank's age, or maybe a couple years younger. She hated thinking something terrible had happened to them. Then again, something horrible had happened to so many people through the years that Nefasylum had been operating, maybe still was happening.
She listened even more closely to Rod.

"The Lieutenant said our squad had been chosen for a special experiment. One that was going to be taking place at a hospital called Nefasylum. None of us had even heard of the place. We were flown out to that hospital. The squad, with our Lieutenant, were taken to the third floor. We were given a room with cots to stay in for the duration of the experiment."

April nodded.
"We all stayed in that one room. The men had to go down to the second floor to use the bathroom, and shower. I had to go to the first floor."
She shook her head.
"Can you believe that?"

Daniel sighed.
"April's right. It wasn't a great place, even before the experiments began. None of us were used to luxurious accommodations, but this place was one of the worst."

Hank frowned.
"So, what were you told about the experiments?"

Rod let out a grunt, and shrugged.
"Not as much as we should have been. It was suppose to be some kind of exercise to help make stronger soldiers. Our squad leader told us to follow directions, and not to ask questions. We were used to doing that. We were taken into a room, where we were told we would be given sensory deprivation. That was supposed to hep us relax so we could learn better. We were placed on these cold tables.

The room was dark enough you couldn't see. As soon as we were in a supine position, and told to close our eyes, suddenly cuffs were clamped onto our wrists, and feet. When we began to struggle, the Lieutenant's voice could be heard coming from some type of speaker telling us to stop fighting."

Shaking her head, April frowned.
"We had been trained to obey his voice. The military is like a cult in so many ways."

Nodding, Rod sighed.
"Sadly, that's fairly accurate. To be put on that table. With no means to fight back, by a person we all trusted, that was a reality check for me. The next thing that was a big wake up call to what we were in for, was the stinging feel on the inside of my arm. I knew someone had just given me an injection."

Shaking his head, Daniel looked at Hank, then over at Missy.
"For me, that was the worst. Going into the service, I had always hated needles. I did my best to get through the string of assorted vaccinations they made you get, then to be in that dark room, and have somebody, I couldn't see, jabbing a needle in my arm, that was almost more than I could take."

April nodded.

"The worst was not knowing what was in that needle and being told not to ask questions. Even if we would have asked, you knew you wouldn't have gotten a truthful answer."

Rod grunted.
"It didn't take long too know it wasn't something good, I can tell you that. Although it was pitch black in that damn room, I could still feel that room spinning. Then the images began floating. I never took drugs, but I'm sure what were given was a psychedelic drug."

A frown line slashed across Missy's forehead.
"You mean something like the LSD, and other drugs, that the MK Ultra program used?"

Rod nodded.
"Exactly something like that. They used that long before our team was used for experiments though."

Missy looked over at April. The woman had her eyes closed. Missy watched the woman's body shake slightly as a shudder ran through her body.

April was shaking her head slowly back and forth. When the brown eyes opened, they stared at Missy, carefully avoiding looking at the men beside her. The men who were once part of her team.

"I can still see those hideous images. I wake up at night trying not to scream. All my life I've been deathly afraid of spiders. Even tiny ones have me hollering out for my husband to come and kill those things. After we were given those shots, the colors I saw were beautiful. They were intensely vivid. They were short lived, though. Replaced by spiders. Awful creepy things. They were all over me. My hands were tied down to that table. I couldn't wipe those things away. That was the worst. With my legs tied down as well, I couldn't move. I didn't want to scream. I just tossed my head back, and forth, forcing my mouth to stay clamped shut. Oh my God, if one of those things would have crawled in to my mouth."

Reaching up, April covered her mouth with both hands.

The sound she made had Missy thinking the woman was about to throw up.
"Are you okay? Do you need a break?"

Swallowing hard, April finally drew in several deep breaths.
"I'm alright. It's just thinking about all of this. Bringing it back into the light of day. Those rotten bastards. How could they have been allowed to do those things?"

Beside April, Daniel sighed.

"For me, it was falling. Even as a kid, I'd have dreams of falling. You now the kind, you are just dropping down from an enormous height. Your body jerks and wakes you up. I used to think if I didn't somehow find a way to make myself wake up, I might have just died in my sleep."
Daniel laughed. "Seems silly to me now. I felt that same way back in that crazy asylum building though. I saw the same beautiful colors, then everything changed. My worst fears hit. I was falling. I was so terrified, I couldn't scream. Believe me, if I would have found my voice, people would have heard my screams miles away. I wanted to reach out, try to grab for some kind of hold to stop the fall, but, like April said, my hands were tied. That was what made me feel more fear."

Rod groaned.
"I don't think any fears I had were that alarming. What I saw after the colors, were hands. Hundreds of hands, reaching out, trying to touch me. I couldn't push them away. To this day, if I'm in a crowd, and accidentally get touched, that same feeling comes back to me. Needless to say, I avoid situations where a crowd is gathered. My wife is a psychologist. She's tried several ways to help me with that. Whatever Neff did up there, goes deep into my psyche. Mine and any others who spent time up in that place."

Missy frowned.
"What about the rest of your team? I'm sure the leader of your squad didn't participate, but you said there were seven of you."

Rod nodded.
"The Lieutenant never was part of the experiments. He was there to make sure we followed orders. Two of the seven, committed suicide, two didn't talk about what happened. None of our team stayed in the military. Not even the Lieutenant. That leaves the three of us from the team."

Missy gasped. "Two killed themselves?"

Nodding, April sighed.
"I'm surprised more didn't. I'm sure many of us had thoughts about trying that. I can't begin to imagine how many people were placed in that asylum over the years. How long was that place open?"

With a grunt, Hank answered.
"Thirty five years. Can you believe that? Nefasylum wasn't forced to close it's doors until two thousand."

The trio exchanged incredulous looks.

Neither Hank, nor Missy brought up their worries that something was still going on up at the building all of them were talking about.
Instead, Hank changed the subject.
"What about you three and the rest of your team? How long did the experiments you endured last?"

Shrugging, April was the one that answered.
"I've thought a lot about that, myself. I know it was at least several days. I told you I had to go down to the first floor to use the bathroom or shower. That's the floor where female patients were housed. I can only remember going down there a few times. In that room, where we were tied down, and filled with drugs, I saw buckets in the corners. Some times they were full, and other times, empty. I'm sure those people had all of us using those buckets to go to the bathroom. Two of our team were women. One of the women, is dead, she was one of those suicides we told you about. To think now, that she and I were going to the bathroom in a bucket with the rest of the team, the weird doctor, and God knows who else, watching, is mortifying. Why would anyone do that?"
April shook her head again, her eyes were moist.

Staring at her, Missy's heart felt like someone had tied it in knots.

"I know this is hard to even begin to talk about. It took a lot of courage for any of you to walk in that front door."

Hank nodded.
"We appreciate you three being willing to tell us anything about all this. Missy and I can't tell you why we feel it's important now to tell this story. I will say that by the time all this was going on, most people thought things like this happening were past history. The types of things that happened, then were stopped in the nineteen fifties or sixties. We want to make sure they never happen again."

Daniel nodded.
"Bringing it out in the open should help."
Then he laughed.
"I guess I'm the wrong one to say that. Besides April, and Rod, I haven't told the story to anyone, except the two of you. I never married so I didn't have to share it with a spouse or kids."

Shaking her head, April sighed.
"I didn't tell my husband for years after we married. I never told anyone in the military. Besides the drugs, I was sexually abused. That was hard to talk about."

Rod nodded. "A lot that happened, feels more like a nightmare, than something that was based in reality.

We were all given shock treatments. We were asked simple questions. Like our names, where we were born, and things like that. If we told the truth, we were shocked, until we could lie and sound truthful."

Nodding, Daniel sighed.
"They used hypnotism on several occasions. Much of that, I can't remember. That scares me more than the other things they did to us."

Drawing in a deep breath, April covered her mouth. She shook her head.
Withdrawing her hand, her eyes were red rimmed.
"I have so many nightmares. I have to think many of them come from those sessions when our team was hypnotized. I think they made us do things to some of the patients."
She began to cry.
"I hope I'm wrong. I dream I killed people, patients, in that place. I know the military has had several secret operations where they trained soldiers to do things they wouldn't normally do. I wonder a lot if that's why we were in that place. The whole thing could have been some type of black operation that was under taken. I can remember flashes of having wires placed on my head, and some types of shocks going through me. Our group could have been used to hurt others, maybe kill them."

She covered her mouth again, as the sobs stopped any other words she had to say.

Again, Missy's heart broke for the three people in front of her, and the others who Neff had been allowed to do such awful things to.
"Would you like to take a break?"

When April shook her head, Rod sighed.
"Actually, I think the three of us have told you what we know. Like April said many of the things that happened were pushed to the deep recesses of our minds."
He turned to look at Hank.
"You said you could write this article without putting our names in it. Does that still hold?"

Hank nodded. "No names will be mentioned in the article, nor anytime in the future. Missy, I, and this newspaper, protect our sources. We consider doing that a sacred obligation."

Rod stood. "I think we've told you all we can, and should go then. Thanks for listening to us, and not acting like the three of us were crazy."

When the other two stood. Hank sighed.
"I've heard more stories about that place, than should be possible. I would never think any of you are crazy. Neff was the crazy one, and someone

should have stopped him before he ever got started doing the things he did."

The trio shook hands with Missy, and Hank then started to leave Hank's office.
He frowned and spoke quickly.
"One more question."
When the three turned back, Hank drew in a breath, "Have you ever hear of a company called RavenWing?"

The three all shook there heads. Rod frowned.
"Why, should we have heard of them?"

Shrugging, Hank sighed.
"No, not really. I'm interested in them, because they are listed as owning the Nefasylum building now. Just curious about that, I guess."

Not sure what to think about Hank's question, or his explanation, the trio left the newspaper office.

After they had gone, Missy turned to stare at Hank, shaking her head.
"This whole thing is crazy. We could make a television series that lasted a decade, or more, about what that asshole did up there."

Hank nodded.

"People would think it was some kind of science fiction series. Instead of based on actual events."

A grunt sounded, Missy nodded.
"You're right about that. Will we be working on this story soon?"

Hank shrugged. "We can take a few days. I have a feeling a lot of people will be coming forward with stories."

Missy sighed. "It's so sad, and so wrong. Nothing like this should have ever happened. We have to make sure it never happens again. No matter what it takes."

Hank stared at her.
"We'll do that Missy. Just don't get ahead of the game. We have to be careful. There's some powerful people who could be behind all of this. We know they were fine allowing Neff to do the things he did."

Missy nodded. "Don't worry. I won't do anything without you, and Luke by my side."

Hank nodded. "Good to hear. Are you going to Luke's for dinner tonight?"

Missy nodded. "Yes, one more night at his place. Then tomorrow at my place. The next night is Halloween. Do you have plans then?"

Hank shook his head.
"Not that I know of. Probably just going over the things we learned from those three today."

Missy nodded. "I'll talk to Luke. I think the three of us should have dinner delivered to his place, and hand out candy to the kids from there."

Hank nodded. "I'd like that."

Missy smiled. "Good, oh, one more thing."

Hank frowned. "What's that?"

The smile broadened.
"Thanks for adding my name to the story about Nefasylum. It means a lot to me."

Looking at her, Hank grinned.
"If you didn't deserve it, I wouldn't have done it. Now, get out to your desk, and give George a break."

Missy did as she was told. She spent the rest of the day writing down some of the things she wanted to remember from the meeting so she could share them with Luke.

After work she headed to his place. She went over everything with him as the two had dinner. Then while they had coffee.

They made plans for dinner at Missy's for the next night, then for Missy, Hank and Luke, to all get together on Halloween night to have some food delivered to Luke's house.

Both Luke, and especially Missy, who loved children, were looking forward to handing out candy to all the ghosts, goblins, and whoever else might come dressed up.
The wonderful sounds of them, yelling out, trick or treat, while they waited for their candy.

Chapter 10

After leaving Luke's Missy drove home. She had planned on just going to bed, but she couldn't get her mind to slow down enough to even try. She kept seeing an image of April, breaking down, and sobbing in the newspaper office. It had been years since April, and the others, had gone through the awful experience at Nefasylum. It would seem that the things Neff did, would never stop hurting. The one thing that bothered her the most was knowing the military had sent people to Nefasylum. Not to look into something wrong, but to be exposed to Neff and his experiments.
The military was known to work with, and have government backing. How did someone fight back against that kind of power, and money?

Missy knew sitting in her apartment, wasn't going to make answers appear, nor make her feel any better, What she needed to do was go take another look around Nefasylum. If she could take look in that window she and Luke had gone to before, she might see more than they had. She could take her phone, get some pictures, then get the hell away from the place. She had told the two men she wouldn't try anything on her own, but they wouldn't have to know. Taking a few pictures would only take a couple minutes.

She didn't even need to use the flash on her phone's camera. When the three of them had gone inside the building she had used the flash, but that wasn't absolutely necessary. She could snap the pictures, then bring them home and edit them. That way, she wouldn't be noticed.

Making up her mind that she was doing the right thing, Missy grabbed a light jacket, her phone, and purse, She locked up her house, got in her car, and drove drove to Nefasylum.
The closer she got to the building fewer, and fewer, street lamps lit the streets.
When Missy parked a block away, she was surprised by how dark it was. She shrugged, hoping that meant she would be hidden from anybody observing her movements. She concentrated on that thought, and not the creepy feeling in the night air.
Getting out of her car, Missy grabbed her purse.
As she walked, she pulled out her phone.
Retracing the walk she had taken with Luke, she made her way to the side of the building.
Seeing the window, with a light on, Missy hurried toward it.
Kneeling down, she frowned at the filthy window pane. Pulling on her jacket sleeve, she used it to wipe a spot large enough so she could see better. She leaned toward the window.
Inside she could see several long tables. To her, they looked like operating tables. To one side she

could see a metal table running along a wall. It held several jars, and various tools. Instruments that Missy couldn't name. She started to lift her phone.

From behind her, a hand reached out, and snatched the phone from out of her hand. Frowning, Missy began to turn her head. Before she even made a half turn to see who had taken the phone, Missy felt something hit the side of her head.

The pain exploded, from her right eye, running down to her jawbone. Missy blinked, seeing pinpoint flashes of light.

She felt something pushed against her mouth tightly. The first image that entered Missy's mind was one of someone using duct tape. Then some kind of cloth was dropped over her head.

She tried to shake her head, then throw her arms around. Anything to stop what was happening. Whoever was behind her, grabbed her hands, and pulled them roughly back behind her. The pain in her shoulders was extreme. To Missy it felt like her arms were being ripped from her sockets.

She tried to yell out, but whatever was over her mouth stopped any screams she would have made. She felt like she couldn't breath. With her mouth taped shut, Missy tried to draw air into her nose. She had to force herself to slow her breathing, while she tried to twist her body.

Moving around, she was able to get into a sitting position. She tried kicking out, but with the cloth over her head, Missy had no idea which direction her legs needed to go. She began thrusting her legs in every direction. A hand gripped one ankle and twisted it. Her captor, whoever it was, hit her knee at the same time.

The sharp, shooting pain, had Missy doubling over. Any thoughts of using her legs for self defense forgotten for the moment.

She felt something, like a rope, being wound around her ankles. Still seated, Missy stayed hunched over, waiting for the pain to ease. Hoping it would. She wondered if her knee cap was broken. Before that pain was gone, her arm was grabbed roughly, in a tight grip. She felt herself being pulled along the ground. Now, she was lying sideways. Her arm, tied behind her, was lifted enough to once again make the shoulder pain return. Her body was half in the air, but her legs and feet were being dragged across the ground.

The only thing Missy could think to be thankful for was that she was being pulled across grass.

That thought didn't last long, as she felt the surface beneath her change to a harder one. She didn't know if she was being pulled along a sidewalk, or some kind of driveway. She could feel the coldness through the legs of her jeans.

A moment later, things got worse.

Her whole body was jolted. Missy knew she was being hauled down steps, most likely made of cement. Her shoulders, then her legs, cried out as her body dropped down each step. In her mind, she felt she could see the dark bruises appearing on her legs.

Finally, the movement down the steps came to an end. Missy felt like she was going to pass out.

The pain more than she could bear.

Behind whatever was securing her mouth, she tried to bite her own tongue. Missy was hoping a new pain, might bring her back from the edge of unconsciousness. She bit down, as hard as her body would allow her to do. Hurting yourself, was something that the mind apparently fought against. Still, her attempt helped, but only slightly,

Then she heard a man's voice.

"Put her on the table in room three. Make sure she is secured. I don't want her finding a way out of here."

Once again, the grip on Missy's arm tightened. She was pulled roughly across a hard floor. A few minutes later, she was lifted up, and thrown onto another hard surface. She knew it had to be on the table the man in the other room had talked about. She hadn't heard her abductor speak, but guessed by the strength, it must be a man, a large man. When she had been lifted to the table, the person didn't seem to have a problem with her weight. He removed the binding on her wrists, and strapped her instead, to the table. Her legs were left, not only tied together, but then also secured to the table, as well. She hadn't heard her abductor walk away, but had heard the sound of a door opening, then closing.

She was now alone, and she presumed, in the dark. Missy focused again on her breathing. It was hard to do with the anxiety she was feeling.

Being captured was hard enough, but she was more afraid of what was ahead of her.

After hearing just some of the stories of the things people had endured in this hideous place, Missy knew she was in big trouble.

She hadn't told either Luke or Hank, that she had headed this way. She also hated the thought of the two men coming to look for her, and ending up in a similar predicament to her own.

On the table, Missy tried to listen. She was hoping to hear any voices coming from the other room. She wasn't having any luck.

She did think she could hear a rustling noise. All she could think of was a mouse, or more likely a large rat, chewing on something.

Missy tried to wipe that image from her mind. She wondered why she could hear a faint noise like that, but not voices. Whatever was making the noise she was hearing had to be in the room with her.

The door to the room had become a barrier, blocking outside noise. That thought didn't improve her outlook. Missy tried not to focus on her pain.

Her whole body ached. The blow to her head, her wrenched shoulders, the wounded knee, not to mention the bumps, and bruises, that had occurred. Besides all that, she felt a headache, like none she ever had, coming on. The pain was intense. Her closed eyes seemed to allow the scrutiny of her pains, bruises, and anxiety more intense.
So much so, she almost missed the sound of the door opening.

Still trying to listen, in order to piece together what was happening, Missy's heart skipped a beat, then the pace quickened until it felt like it would pop out of her chest. Then someone touched her. Tied up, she couldn't pull away.
She felt the covering on her head lift slightly. The material was moved just so her mouth was in the open air. The tape on her mouth was ripped off. Missy felt part of her lips come away with it.
She screamed.

With the hood, or whatever it was still over her eyes, Missy couldn't see who was in the room with her, but she could hear the sound of a male, laughing.
"Scream all you want. No one will hear you."

Missy drew in several deep breaths. She let the air fill her lungs. Relieved to finally be able to breath normally again. She wished she could see the face

of the man talking to her. She ran her tongue, sore from the bite she had given it earlier, over her dry, tender lips.
"Who the hell are you? Why have you brought me here?"

The man laughed again.
"You're mistaken. I didn't bring you here. You were trespassing. This building is private property. You were not given permission to be here. The question is, not why did I bring you here, but why were you trespassing? Why were you trying to take pictures of this place?"

Missy shook her head.
"I was just curious. No reason that I can give you other than that. Even if you feel I was trespassing, why not just call the authorities? You tied me up, knocked me around, threw me in this godforsaken place, and stole my phone, and apparently my purse, as well."

A moment of silence, then the man spoke again.
"Melissa Carlyle, twenty eight years old. You live right here in Mansfield, and if I'm not mistaken, work for the Mansfield Blotter. Does your job at the paper, have something to do with your trespassing, and this curiosity you spoke about?"

On the table, Missy tried, fruitlessly to struggle against her restraints. She could feel her anger growing.
"I have nothing to say to you. Just take off these bindings, and let me go."

Under the cloth covering, Missy didn't see the raised hand. She felt it as it struck her face, though.
The man's voice was spoken slowly.
The words clipped.
"Perhaps you need time to think about not talking to me a bit longer."

Missy heard the sound of the door, opening and then closing. She felt like crying. Her eyes burned with tears that didn't fall. She strengthened her resolve to not let herself give into tears. That was the last thing she needed. Despair at her situation, would do her no good.
She needed to clear her mind and think.
Missy was finding that hard to do.
Pain was over riding her thoughts.
Every bone, muscle, and nerve, in her body seemed to be signaling to her brain that she needed to sleep to get away from the pain.
Missy was certain that screaming would do her no good. She had no doubt the man had been telling her the truth about that.

Right now though, she had no idea what she could do. Time seemed to stand still as she listened for the door to open again.

She dreaded hearing the sound, yet, almost prayed for it to happen.

Instead of trying to clear her mind, Missy began singing to herself. Then reciting poems.

Anything she could think of to make time pass more quickly.

She laughed quietly, wondering why doing that even mattered.

Chapter 11

Luke was sitting in the kitchen when the phone rang. He answered it, with a quick hello, then frowned hearing Hank speaking.

"Have you seen Missy?"

Luke frowned. "Not since she left after dinner last night. Isn't she at work?"

Hank sighed. "No, I've tried calling her, and her phone goes straight to voicemail. That never happens. She's been walking to work, and should have been here by now. Did she say anything to you about talking with the three people from the military yesterday?"

Luke nodded. "Yes, I know she was upset about it. I can understand how she feels, Thinking that the government, and the military, would allow Neff to do experiments on our own service people is beyond belief."

Hank sighed. "I'm going to head over to her place. I need to find out where she is. I'll have George watch the office."

Luke groaned. "Oh hell, would you be willing to stop by my place, and pick me up? I hope Missy didn't get hurt walking to work. I know she doesn't drive unless it's snowing or something."

Hanks nodded. "You're right about her driving. Just let me tell George what's going on, then I'll swing by your house."

Feeling his anxiety growing, Luke nodded. "Thanks Hank, I'll be waiting out front."

Grabbing his coat, Luke hurried outside.
He knew a lot of different things could have happened to Missy. Numerous reasons for why she hadn't made it to work. The one thing he tried not to think about, was Missy heading to Nefasylum on her own.
Luke paced up and down his sidewalk, until he saw Hank coming up the road.
He ran to the man's vehicle, and jumped in.
"Thanks for swinging by here."

Hanks nodded. "No problem, I know how close you and Missy have become. I'm going to drive by the newspaper office, then make my way down the route Missy usually takes when she walks to work. Keep an eye out for her, or for her car."

Luke nodded. "Sure thing. If she's on the road somewhere I'll see her."

While Hank drove, he was also scanning the area, hoping to find Missy standing along the way.
As each block went by, he felt his heart and hopes dropping. Making his way to Missy's apartment, Hank drove around back where he knew she always parked her car. He let out a groan.
"Oh hell, her car's gone."

Luke shook his head. "Now what?"

Staring at Luke, Hank shrugged.
"Now, it's time to drive toward Nefasylum, and see if we can find her car."
Driving up the road, Hank hit the steering wheel hard with his hand a couple times.
"Damn her, Missy told me she wouldn't go up there alone."

Luke sighed. "Maybe she didn't. She might have gone shopping before work, or something. She might have had a flat tire or other problem like that."

Glancing over at Luke, Hank grunted.
"You don't really believe that do you?"

Luke sighed and shook his head.

"No, I wish I did."

Lifting a hand, Hank pointed to the side of the road. The two were about a block from the Nefasylum building.
"There it is. That's Missy's car. Let's go have a look."

Luke nodded, but didn't speak. He felt like he'd lost the ability. When Hank parked behind Missy's car, both men got out of Hank's SUV, and walked to Missy's vehicle.

Hank shook his head.
"Her car hasn't been hurt. That means she went up there, and alone. Probably after she left your place last night."

Luke frowned, turning to Hank.
"What do you think we should do? We have no idea how many people might be up in that building. Oh hell, I hate to think of Missy up there with scum like that."

Nodding, Hank's eyes narrowed.
"I know a guy that works for the Mansfield police department. I hate to leave Missy inside that building any longer, but first we need some help to get her out."

Staring at Hank, Luke frowned.
"Do you think this guy will believe that someone from Nefasylum took her?"

Drawing in, then letting out a deep breath, Hank shrugged.
"I hope to convince him of that, and get more cops to come out here as well. Right now, the best thing we can do for Missy is go and talk with the police. Convince them to get up here."

Luke nodded his agreement. He knew what the people who worked for a place like Nefasylum were capable of. He wished he had a way to fight them himself, and get Missy back. He was also a realist, and knew Hank's idea was their only option right now. He only hoped the local police hadn't been corrupted by those running Nefasylum.

Pulling in front of the police station, Hank turned to Luke. He let out a sigh.
"Come on. It might take both of us to talk the cops into going up there."

Walking into the station, Hank stepped over to the main desk. Luke stayed close to him.

A receptionist, who was also the station's dispatcher looked up at the men.
"Can I help you?"

Hank nodded. "I'm looking for Spencer Zauner. I'd like to talk to him."

The woman's eyes narrowed slightly.
"Can I get your name?"

When Hank gave it, she picked up the phone, and talked quietly in it.
A moment later, she looked back up.
"Deputy Zauner said to send you back. Are the two of you together?"

Hank nodded. "Yes, this is my friend Luke. We won't take much of Spencer's time."

The woman nodded, but her eyes remained narrowed with skepticism.
"Do you know the way to his office?"

Hank nodded. "I do, thanks."

The two men hurried away from the desk, and down to an office further inside the building.
Just as Hank raised his hand to knock on a door, it was pulled open.
Blue eyes stared at Hank, the sparkle in them, held a bit of humor.
"Good morning Hank. Looking for some juicy tidbits for your paper?"

Hank rolled his eyes.
"Nope, I have enough to keep me busy for a lifetime. I do have a problem though. Do you have a moment to talk?"

Pulling the door open wider, Spencer nodded. Then he frowned seeing Hank wasn't alone. Noticing the look, Hank smiled.
"This is my friend, Luke. Don't worry, he isn't dangerous."

Reaching out, Spencer shook Luke's hand.
"Hi Luke, I'm Spencer, Deputy Zauner to dangerous people. Both of you sit down, and tell me what's going on."

As the men took seats, Spencer moved behind his desk. Hank drew in a deep breath, wondering how to tell Spencer what he thought was happening with Missy.
"Have you had a chance to look at the article the paper put out on Nefasylum?"

Spencer nodded. "I did. It's about time someone told the truth about that damn place."

Hank nodded. "That article has only scratched the surface. We've had calls from others who know about what happened up there, and are willing to talk about it. You remember Melissa Carlyle from the paper?"

A smile, was followed by a nod.
"Sure, I know Missy, she's great. I'm surprised she puts up with you. She should be the number one writer at the paper."

Despite his worries, Hank's chuckle filled the room.
"She's told me that herself. I'm scared she's in trouble now."
Hank pointed at Luke.
"Missy and Luke saw lights in the basement up there. The two of them looked in the windows, and saw something is going on again in that building. I even asked Owen Carter to take Missy, Luke and I, up into the building. He's the security guard up working there. He didn't have access to the basement. It was locked up. The power isn't even on anywhere else in that building."
Drawing in a breath, Hank sighed.
"Missy drove up there sometime last night, on her own. Now, she's missing. She doesn't answer her phone. It goes straight to voicemail. Luke, and I, found her car. It's parked a block away from the building. I know they took her. If I'm wrong, I'd be happy, but I just want to rule out what I think happened. The only way to do that, is by going up there. Luke and I don't have enough manpower on

our own to do it. Would you trust what I'm saying enough to go to Nefasylum, and help us go get Missy?"

Sitting back in his chair, Spencer stared at the two men sitting in his office. The genuine worry on their faces was obvious. Finally he nodded.
"I won't say I would normally trust you. The two of us have known each other for a long time. I believe what you're saying. Or believe you feel what you're thinking is true. I'll get my partner, and a few other cops, and we'll head up there, and take a look around. Do you have the security guards' phone number? If he could get us into the main building, we can have a better chance of surprising whoever is in that place. Could you call him?"
When Hank nodded, Spencer shook his head.
"What the hell is someone doing up in that place anyway? It should have been condemned and torn down years ago."

It was Luke who answered.
"Whatever they are doing, with their track record, you know it isn't something good."

Hank, and Spencer nodded at that.
Spencer turned to Hank.
"I should have a team together within a half an hour. You get a hold of this security guard. Tell him to meet us up where you said Missy's car was. If he

says no, give me a call. I'll make sure he agrees to do it. If he consents, then we'll see you up by Nefasylum."

Thanking Spencer, Hank, and Luke left the police station. As soon as they got outside, Hank called Owen Carter. It took a few moments of convincing, and invoking the deputy's name, but Hank finally got the man to agree to meet them.

The two men got in Hank's SUV and drove back to Missy's car. Looking ahead at the vehicle, Luke was praying that magically Missy would just be sitting inside. Although he knew that was impossible he still felt the hope draining out of him, like the air coming from a balloon hit by a sharp instrument.

Getting out of his door, Hank motioned up the road. "Here comes the cavalry."

Turning, Luke grinned slightly, seeing three police cars headed towards them.
Behind the men, someone cleared their throat.

They turned to see Owen Carter standing there. "Hell, I hope you're not going to get me in trouble. I'm just the damn security guard. All I want is to walk the floor, draw my wages, and mind my own business."

Hank grunted. "You're not in trouble Owen. All the police want is for you to unlock the front door of Nefasylum, then you can just walk away. If things go like I hope, you might be out of a job anyway. If I get what I want, and accomplish what I know is the right thing to do, this place will be bulldozed to the ground."

Owen stared at Hank.
"Fine, I'll unlock the door, but I ain't waiting around for those cops to come talk to me."

Luke was grinning, while Hank nodded.
"Okay by me. Let's go unlock it."
He turned to Luke.
"Will you wait here, and let Spencer know what we're doing?"

Luke nodded. "Go on, I'll talk to the deputy."

Hank, and Owen walked away. Luke stepped to the back of Missy's car, waiting for the police to pull in and park.
As soon as they did, he brought Spencer up to date. Telling him what Hank, and the security guard, were doing.
Nodding, Spencer walked to his team, and gave orders for what he wanted them to do. He sent two deputies to watch the only basement door. When he

gave a signal, they were supposed to break down the door, and come in. Another deputy watched the perimeter, around the main floor of the building. Spencer and his partner, Denise Trippet, were the ones who would be heading into the main building of Nefasylum. They planned to go through the door Owen Carter had unlocked, before he hurriedly left the area. While Spencer gave instructions, Hank returned,

Spencer and Denise stepped over to him.
Hank was now standing with Luke.
Spencer stared at both men.
He pointed at the red haired woman standing beside him. The hair was pulled back tightly. A hat, part of the uniform, sat on her head.
"This is my partner, Denise Trippet."
Spencer looked at Denise.
"Meet Hank. and Luke, the two men that have us standing out in the cold this morning."

Green eyes smiled, as the woman grinned.
"Ignore him. I'm happy to meet you both, and hope we can find your friend."

Clearing his throat, Spencer spoke to Hank, and Luke.
"I want you both to stay outside this building. I mean that. Denise and I will go in, and head down to the basement."

Hank frowned. "I got the security guard to unlock the front door. Then he hightailed it out of here. The door to the basement has a padlock on it though."

Spencer grinned. "I have something to fix that." He turned, and headed toward the police car.

Staring at the man, walking away, Luke wondered what tool Spencer had in mind..Then he frowned, and spoke to Denise.
"Better grab some flashlights also. The main floors of that building don't have power. Just the basement has the electricity turned on. You have to walk to the far end of the main floor to get to the doors to the basement."

Looking at Luke, Denise slapped her hip.
Green eyes were shining with humor.
"No worries there, I always have my trusty flashlight with me."

Returning to the others, Spencer held up a pair of bolt cutters.
"These should take care of the padlock. The doors leading to the basement, Denise and I should be able to shoulder our way through. An old building like that shouldn't take too much. Like I said before, I want the two of you to stay here. No

matter what you hear, or even see. I don't want to have to babysit the two of you."

Hank grunted out a laugh,
"We'll stay put. Just go find Missy."

Nodding, Spencer, and Denise headed to the building.

A few minutes after the two were gone, Luke turned to face Hank.
"I know we should stay here by the car, but do you think we could at least get a bit closer to the building? Hell, we're almost a block away."

Hank nodded. "I'm with you, let's move closer. We just need to stay where no one can see us."

By the time Hank, and Luke started walking, Denise, and Spencer had just opened the door leading into the main floor of Nefasylum. They turned on their flashlights, and stepped into the building. The two walked through the reception area and through the doors that led to the patient rooms. Denise turned to Spencer.
She pointed her flashlight down the hallway.
"Luke said the door to the basement was at the far end of the hall on this floor."

Spencer nodded. "Let's go. Just keep an eye out. I'm sure what ever is going on in this crazy place is happening in the basement, but you never know."

The partners made their way down the hall, and found the doors. Spencer used the bolt cutters to cut the padlock. He tried the knobs on the door first, then he drove his shoulder against the double doors. They didn't offer much resistance, flying open. The clunking noise had Spencer reaching out to stop the doors from repeating the motion.
He swore under his breath before turning to Denise. "Okay, we go down. I'll take the lead. Let me tell the others, before we move."
Pushing a button on the communication device sitting on the front of his uniform, Spencer spoke quietly.
"This is Spencer. Do you copy."

A voice came back.
"Ten four, we copy."

Spencer let out a sigh of relief. He had been worried the radios wouldn't work in the building. "Denise, and I are in the building, preparing to head to the basement. One of us will signal you when we want you to break in."

The voice sounded again.
"Copy that. We'll be ready."

Motioning to Denise, Spencer pulled out his pistol, and whispered.
"I'll take the right side, you take the left. Be careful."

Denise nodded, removing her own gun from its' holster. Both the deputies walked down the stairs, slowly making their way to the basement, neither sure what to expect.
Reaching the bottom of the stairs, the two stood in a small foyer. A single door in front of them.
Spencer pointed at it, but stared at Denise, who nodded. Silently acknowledging she was ready.
Both lifted their guns, as Spencer pulled the door open quickly.
The two saw a man standing near a table.
Spencer yelled out.
"Hold it right there, Hands up."

Surprised, the man stared at the two intruders.
He stepped back from the table. Both Denise, and Spencer stepped into the room.
Again Spencer was the one to speak.
"On your knees, hands above your head, now."

At the same time the man dropped to his knees, Denise felt herself shoved to the ground.
Landing hard, she turned quickly, lifting her gun.

She pointed at a man, who looked like he came just came off a football field, and still had on his shoulder pads.
Pulling the trigger, Denise shot the man in the leg, and struggled to her feet, as he dropped to the ground. The sound reverberated off the walls.

Staring at Denise, while still trying to watch the other man, now on his knees as well, Spencer spoke into his radio.
"Get in here. Break the damn door down."

Moving to the man he had spoken to, Spencer pulled the man's arms behind his back, and placed handcuffs on him.
"Where's Melissa Carlyle?"

The man stared up at Spencer, dark eyes filled with hate. His voice when he spoke was angry.
"I don't know who that is, or what the hell you two are doing. What's going on here? You have no right to break into this place. You just shot an innocent man."

A few feet from Spencer, Denise was trying to place handcuffs on the larger man. He was struggling, waving his arms around trying to make contact. Lifting her leg, Denise kicked the man in the same knee she had just put a bullet in.

He let out a scream. While the man was in pain, Denise was finally able to get handcuffs on him. She pushed him back.
"Just sit right there and don't move. If you do, I swear I'll shoot you again."

Hearing Denise, Spencer tried not to grin. She looked like a tiny lady, but he knew better, and had a feeling if the big man tried anything, he would learn that same thing, the hard way.
Spencer grabbed the man kneeling near him by the shoulder.
"I asked you where Melissa is?"

Before the man could answer, the others on Spencer's team ran into the room.
Spencer looked at them.
"Go check around. I don't know how many rooms are down here, but check every nook and cranny. Missy has to be here. Be careful, I don't know if others are down here. We found these two, but haven't checked any where else."

Nodding, the others turned, and guns in hand, began the search. Denise and Spencer were able to move the two handcuffed men off to one corner of the room. Spencer knew he should have someone get to the patrol car, and call for an ambulance, but he wanted to see what kind of shape Missy was in, if she was even down in the basement.

A voice from somewhere toward the back of the basement yelled out.
"I got a guy in this room."

Another voice sounded.
"There's a woman in here."

Wanting to get back to where the voices had called from, Spencer looked at Denise.
"You okay watching these two?"

Denise nodded. "I'll be fine. If they try anything, they won't be though."
She turned and glared at the two, then looked at Spencer, with a grin, and nodded again.
"Go on, sounds like people need help back there."

Nodding, Spencer left the room. He still had his gun out, unsure of what he was heading to find. Walking down a hallway, he saw one of his deputies standing outside an open door.
He motioned at Spencer.
"He's in here. This guy looks like he's been through hell. He had a hood over his head, and was tied up, when I found him."

Nodding, Spencer stepped in the room. A man maybe in his thirties, was sitting in the corner of the room. His face was bruised, and had several cuts

visible. The man looked too thin to even be alive. Spencer moved over to the man, and squatted down, so the two were on eye level.
"We're going to get you some help. Are you up to answering a couple questions?"

When the man nodded, Spencer turned back to his deputy.
"Put a call into dispatch. You'll have to go out to a squad car. Tell them we need as many ambulances as they can get up here. I have a gunshot wound out in the front, and at least two people needing transported back here."

The deputy nodded, and hurried off to make the call.

Beside the man, Spencer stared at what looked like the shell of a man.
"What's your name, and how long have you been here?"

The man licked cracked lips.
"Frank..."
The man began coughing. When he recovered, his voice was weak.
"Frank Harmon. I'm not sure how long I've been here."

Spencer nodded. "That's okay. We'll figure everything out when we get you out of this hell hole. I need to check on others that might be in here. Are you okay until someone gets back in here?"

The man gave Spencer a weak smile.
"I made it this long. I'll be okay."

Nodding again, Spencer stood and left the room. He walked further into the basement. He found several rooms on both sides of the hallway. Spencer looked in the first two, but didn't see anyone. The third room he looked in, another deputy was standing by a bed, removing the restraints that were holding a woman down on the narrow bed.
The deputy was frowning.
"She's confused. There was a hood covering her head. Those assholes kept her in the dark, and disoriented. I just about have these restraints off of her."

Nodding at his deputy, Spencer stepped over.
He recognized the blue eyes, but not much else on the face he saw.
"Missy, it's Spencer. Hank and Luke sent me to come get you."

Turning her head, Missy stared, not comprehending the words she was hearing. Spencer repeated them.

Missy let out a sigh, as the tears rolled down her swollen cheeks. Her eyes had bruises around them. "Thank you."

The words came out slowly, and quietly.
Spencer could hear the raw emotion in them. He reached out and carefully touched her shoulder.
"We called for an ambulance. It won't be long. Things are going to be okay. You're safe now."

The blue eyes closed.
The deputy turned to Spencer, he shook his head.
"I can't believe what we are finding in here. How could this even be going on?"

Shaking his head, Spencer sighed.
"I wish I had a good answer for that."
He pointed at Missy.
"Will you sit with her, while I check the rest of the basement?"

The deputy nodded, and Spencer left the room. He would have rather stayed with Missy, but he was the leader of the team. It was his job to keep looking. He went toward the end of the hall, finding no one else. Then just as he got close to the last door, he heard a moaning noise.
Stepping in, he saw a woman tied down, similar to how the other deputy had found Missy.

Spencer moved over to the woman, carefully pulling the covering from her head.
He saw the scared look in her dark eyes, as she stared up at him.
Spencer spoke quietly, hoping his voice sounded calm. He could see dried blood caked in the dark hair. Like Missy's, the woman's face looked swollen, and bruised. He removed the restraints as he spoke.
"You're okay now. We're here to get you out of this place. We're going to get you some medical attention. Help is on the way. I want you to relax. and stay on the bed. I don't want to move you. I'm afraid to make your injuries worse. The medics will be in soon. I have to go out of the room, but I'll send someone back to wait with you. Is that alright?"
A sigh was followed by a barely discernible nod. Spencer decided he would wait to even try and question the woman. She looked too frail.

Leaving the room, he walked down the hall, stopping at the other rooms, to tell his deputies to stay with the people they had found. Then he went out to where Denise was standing, holding her gun on the two men in handcuffs. Spencer noticed she had found something to use as a tourniquet on the man's leg she had shot. The man was still groaning though. Seeing Spencer, Denise frowned.
"Did you find Missy?"

Spencer nodded. "We did, we also found two others, a man and a woman. I wondered if you would go and sit with the woman. She's at the far end of the hall back there. I'll watch these two."

As Denise began nodding, the sounds of sirens could be heard. She let out a relieved sigh, and headed to find the woman.

Outside the building, Luke and Hank had stopped the deputy, who had come out to make the call, long enough to learn that people had been found in the basement. He didn't know any names, but told the pair that two men had been handcuffed, one of whom had been shot. Then he added that at least two people had been found in the rooms.

Hank and Luke had moved even closer to the building where they knew the outside basement entrance was. Both men waited anxiously for more news. They could hear the sirens coming closer. When two ambulances pulled in, the medics wheeled several gurneys into the basement. Neither man spoke, but both were having similar feelings. The tightened nerves, the difficulty breathing, the evident worry, as the two waited to see if Missy would be brought back out. Watching what was happening, Hank pulled out his phone, and took pictures, knowing he needed to document what was occurring.

After what felt like hours, the two saw the dark hair, they recognized on a gurney. They moved closer. Missy turned her head slightly.
The two men tried not to show their shock. The swollen face didn't look like the one both were used to seeing.
Luke moved to the gurney. He smiled.
"Good to see you. I was getting a bit worried."

Moving to stand beside Luke, Hank nodded.
"Same here. I should tell you how angry I am at you for coming to this place alone. That can wait until after you get a bit of medical attention."

Blue eyes filled with tears.
"Thanks for sending Spencer. He told me you both asked him to find me."
The two men smiled, then stepped back, so the gurney could be placed in the ambulance.

They waited to talk to Spencer. Although they needed to get the hospital, the two also wanted to learn what had happened in the basement of Nefasylum. They watched three people taken out who had been captives, then the two men who had done the damage. Both wondered how many others should be put in handcuffs as well. Those who set everything up, and others. who just weren't in the basement when the cops had arrived.

Now, the problem of finding the truth was with the local office. Maybe even law enforcement a bit higher up rungs of the ladder. The two worried that a cover up would happen again.

Luke mentioned that thought to Hank.

The other man shook his head.
"I'm not letting that happen. If I have to write a thousand articles about this place, and the terrible people here, I will. I also know, as soon as Missy does some healing, she'll tell you the same thing."

After the men talked with Spencer, hearing the replay of what had gone on, they headed to the hospital. Eager to see how Missy was doing.

The two talked as Hank drove. They spoke in disbelief that three people had been found in that place.

Luke was shaking his head.
"I wonder if others were there before, and what happened to them? I told you that my friend Mason noticed that light in the basement a while back, long before he died. Several people could have been in that place, then moved out of Nefasylum, in the

time since. Missy should have never gone up there alone, but at least now the authorities know something is going on."

Hank sighed. "They know. I only hope someone with money, and power, doesn't find a way to hide what happened. Like they have in the past. Like I said, I won't stop writing about it, and hope that will help. I also have faith in Spencer. He's a good cop, and a great guy. He won't like any one trying to stop him from investigating all this."

Luke smiled. "That's good to hear. Fighting against the bureaucracy is never easy. Life is never easy or fair, I wish it was."

Chapter 12

At the hospital, Missy was taken into the emergency room, as were the two others who had been held captive.

Spencer had Denise take the man, the one she hadn't shot, to the jail.
He stayed with the man who had the gunshot wound. Once the bullet was removed, and the man deemed okay to leave, Spencer took him to the jail. After making sure the men had been read their Miranda rights, Spencer and Denise tried to ask questions.
Before they got much farther than asking for names, both men asked for a lawyer, and the questioning was stopped.

Missy was taken to a room to spend the night, She was lucky to not have more serious injuries.
She did have a concussion, a lot of cuts and bruises, especially to her face. Missy had been told the night's stay was just for observation purposes.
Still, she just wanted to go home.
The door to her room was open, but she still heard a knock.
Looking over, she saw Luke and Hank, and smiled. Despite how bad the motion hurt, for Missy it felt good.

"Get in here, both of you. Come on, and sit down. Pay no attention to the way my face looks."

The two men found chairs, and moved them, one on either side of Missy's bed.
Luke laughed. "I'm just glad you're alive. You're lucky, I have a feeling the people in charge of Nefasylum didn't like anyway finding out about what's been going on up there."

Hank nodded. "Luke's right about both things. It's good to see you, and know you're safe. I get the feeling you have more than a few aches and pains. though."

Missy grinned. "They gave me pain medication. It doesn't feel so bad now. I am a bit tired. I should be home making Luke dinner. Sorry about that, I know It's my turn."
Missy frowned.
"That's right isn't it? I wasn't in that place longer than a few hours was I?"

Luke frowned. "Depends on when you got the thought in your head to go up there, and alone."
The frown was replaced by a grin. Luke couldn't stay mad at Missy. He was too glad to see her.
"If you went up there not long after you left my place, then you were in there sixteen hours, give or take."

Staring in disbelief, Missy sighed.
"That Long? Never mind, we can still get together for Halloween. The doctors said I just have to stay tonight. I can hand out candy, and look like I have a strange mask on."
She frowned. "Someone found my purse, the glass on the front of my phone is all broken up. The whole thing looks like a monster crushed it in their hands."

Both men laughed at her mask comment, then shook there heads at her statement about her phone. Seeing Missy's eyes drooping, the two decided the talking could wait. They decided to let Missy rest. Telling her to sleep, and get better, they left.

When Hank drove Luke home, the two made plans for getting Missy home the next day, and then for the Halloween she still wanted to be a part of. Afterwards, Hank headed to the newspaper office to fill George in on the mornings' events.

Missy slept through the night, and even an hour or two later than she normally would have.

Hank had picked up Luke, and the two of them arrived at the hospital to take Missy home. Then took her to her place. Both men walked in her apartment with her.

Missy wanted to go to work, but Hank had given her a firm, no, on her request.
He had gotten Missy's keys, and asked a friend to drive her car home for her.
"You stay home today. We'll get together tonight, and see how you feel. If I think you're doing okay, you can come back to work tomorrow."

Luke nodded. "Hank's right. You still have a lot of aches, and pains. Not to mention the mental anguish. I want to hear the whole story tonight when we get together. That is if you feel up to sharing it."

Missy smiled. "I'll be glad to share the story. Do I need to go buy candy?"

Both Hank, and Luke shook there heads.
Hank laughed. "We both got candy, and we'll order in the food, so the bases are covered. Do you want a ride tonight, or should I come pick you up?"

Missy smiled. "Thanks for the offer, but I'll drive. Thanks for getting my car back here."

Hank nodded. "You're welcome. I need to get down to the office."
He turned to Luke.
"You want a ride home, or you staying here a while?"

Luke sighed. "I'd like to stay and visit, but better go home."
He smiled at Missy. "See you tonight."

She nodded, and told both men good bye.

Just before it got dark, Missy got in her car, and drove to Luke's house. She smiled, seeing he had put up decorations for the holiday. Walking to his door, she could feel the pain in her knee. She had decided to go without the pain medication, since she was driving. She knocked on the door.

When Luke answered, holding a bowl of candy, he frowned at the person on his porch. He didn't think the person in front of him was young enough to be a trick or treater.

Missy was smiling. She had worked for hours on her costume. With her face swelled up, she barely recognized her own image in the mirror. She had covered her face, bruises and all with white make-up. She then had painted on a clown face. She wore sweat pants. Missy had thought about placing a pillow in them, but decided it was too awkward. Instead, she had gone to town long enough to find a clown outfit. She had also picked up a new phone. Tonight, Missy wore the brightly colored shirt, that had been a part of the costume.

Holding out a hand she spoke.
"Trick or treat. Give me some candy."

Light blue eyes widened, then narrowed.
"Oh my word, Missy, is that you?"

Nodding, she began laughing.
"I told you I was going to wear a mask. I decided to just paint my face instead. From the look on your face, I did a good job."

Luke also laughed. "That you did. Until I heard your voice, I didn't know it was you. Come on in. It's cold out here."

Stepping in the house, Missy cold feel the difference in the air. She could tell that Luke had a fire going in his wood stove. She walked over to it, and put out her hands. She rubbed them together. "I wish I had a wood stove. This is great."

Luke nodded. "I love mine, come and sit down. I thought we'd have our food in here, where it's comfy and warm."

Making her way to a chair, Missy tried not to favor one leg. She didn't want Luke to see it was bothering her. She knew her attempts had failed when Luke spoke to her.

"With all the injuries you received, I forgot about your knee. Is sitting in that chair going to be alright for you? I can grab a taller chair so it isn't so hard to get up and down."

Missy sighed. "This is fine, but I don't know if I can hand out candy to the kids, or for how long. I'd like to try. I love seeing the little ones in their costumes."

Luke laughed. "And they'll love seeing yours."

When the doorbell rang, Luke turned to Missy. "Time to find out how good that knee works." He stepped over, and held out his hand.

Missy grabbed it and used it to help her get up. She smiled.
"Not too bad, thanks for the help."

She went to the door, grabbed the bowl of candy, and stared at three little children. The faces of an angel, a super hero, and a cat, smiled, staring up at the clown, who had answered the door.
They all yelled out.
"Trick or treat."

Giving them candy, Missy closed the door, and just got back seated when the sound of the doorbell was heard again. Luke laughed.

"Let me handle this one."

Luke made more trips to the door than Missy, but she was still happy at the children she was able to see. Missy loved kids, and hoped some day to have her own.

In between trips, Luke's phone rang.
It was Hank calling.
"Hey, I thought since I'm in town, I'd grab some fast food for us. I'm pulling into Julie's. The fast food place on main. What do the two of you want to eat?"

Luke relayed what Hank was doing, to Missy.
She smiled, knowing the place, and nodded.
"I'll have the chicken tender basket, and a strawberry shake. I don't know if I can eat the chicken right now, but the fries should be soft, and a shake would be great. My jaw is still sore."

Staring at her, Luke hated to see Missy hurting. He hoped the physical pain would ease soon. He knew the mental anguish would take much longer to heal, if it ever did. He nodded at her, and told Hank what Missy wanted, then asked for a cheeseburger, fries, and a chocolate shake, for himself.

On the other end of the line, Hank nodded.
"Got it. I'll be there soon."

Hanging up, Luke turned to Missy.
"I'm sorry for the pain you're going through."

Missy sighed. "A lot of people, including yourself, went through much worse. Having suffered just a bit of what others have, that might make the articles I help Hank write more realistic, and empathetic."

Luke nodded, but still felt terrible for Missy.
While the two waited for Hank, and the food, they handed out candy, and talked.
For now, neither spoke about anything related to Nefasylum, or Missy's treatment while being help captive in the place.
Both knew the time for that was coming later.

Once Hank arrived, the first thing he did, was stare at Missy, and then break out laughing.
"You look different. I almost didn't recognize you. I have to say even with all that on your face, you're still the cute Missy, we all know and love. Now, let's eat. I'm starving."

The three sat down and ate. Missy didn't eat her chicken, but was able to eat the fries, and enjoy the shake. The trio took turns handing out candy. When the sound of the doorbell ringing finally tapered off. All if them knew the time to hear Missy's story had come.

When Hank had brought in the food, he had also carried a folder. Now, he lifted it, and drew out some newspapers.
He handed one to Missy, and another to Luke, while keeping one for himself.
"I didn't want to show you this until after we ate. I also wanted Missy to see it, before she told us about her being in Nefasylum. That is, if she is up to doing that."
Luke, and Missy stared at the paper. It was dated for the following day. Hank smiled.
"That's the front page for tomorrow's paper."

Holding the paper, staring at the picture on the front, Missy's hand trembled slightly. The image showed the back of the Nefasylum building.
Where the door led to the basement, ambulances were lined up. Someone was being brought out on a stretcher. Although a face couldn't quite be seen, by the size, and shape, on the stretcher, Missy was sure who she was looking at. The person being brought out had to be the man who had abducted her. The man who had caused most of the pain she was going through now.
She sighed, and looked at the paper's headline.
She read it out loud.
"Is the horror of Nefasylum finally over?"

Missy and Luke read the article. It stated that two men had been arrested. They had been found in the basement of the building with three captives. No names were given in the article, but the conditions of those who had been abducted was told.
Both looked up when they had finished reading the entire article.

Hank stared at the two in front of him.
"I left out any names. I do know the man's name who was found, but not the woman's. I thought it was better this way."

A sigh came from Missy.
"I like it better that way."

Both Hank and Luke were staring at her. It was Luke who posed the question, both men were thinking.
"Do you feel up to telling us what happened in there Missy?"

She nodded, setting her paper on the coffee table. Missy drew in one breath, then another.
"I wasn't planning on going up there without the two of you. I couldn't stop thinking about that place, after those three from the military came to the paper. I just thought if I could capture a picture or two of what was happening in that basement, we would know what was going on. It was a mistake,

but at least it brought the police to Nefasylum. I only hope, the military, or the government, don't manage to cover it all up again."

She drew in another breath, and closed her eyes trying to recall the events. Missy's eyes remained closed as she began speaking.

"I was outside the basement window, preparing to take some pictures. I lifted my camera, and someone grabbed it from my hand. I never saw his face, nor the other guy's."

The dark blue eyes opened. Even behind all the makeup, both men could see the anguish.

"The other man was behind me, he hit me so hard, I dropped over to the ground. Some kind of tape was slapped on my mouth. Sometime during all that, he hit my knee hard, I thought it was broken. The guy slipped something over my head, tied me up, and pulled me across the ground. Then down into the basement. Those stairs are hard, and felt like I was dropping a few feet with each bump. I was taken to a room, and strapped on a table. I lost track of time. I'm probably not getting things in order. I couldn't see, and the pain was almost unbearable. The other man came in my room. He lifted the hood, or whatever it was, just so my mouth was showing. He ripped the tape off. That hurt like hell. He took half my lips when he took the tape."

Reaching out, Missy grabbed her shake, and took a drink, before continuing.

"Some time later, the second man returned. I don't remember exactly what he said, something about this won't hurt, then you'll be out of pain. I felt a stinging sensation on my elbow. Whatever he injected, felt warm. A few seconds later, everything changed. It was like having a dream, or more like a nightmare. Images floating through my head. Awful things, spiders, snakes, rats, other stuff I can't remember. In the background the man asking questions. I don't remember if I answered or not. I'm sure I did. I probably spilled my guts about every thing that ever happened in my life. Whatever I told him, he wasn't satisfied with, because he slapped me several more times. Then pulled the hood back down over my mouth. I was just happy he didn't put the tape back on. I was trying to figure out how to get out of the place. I couldn't seem to keep a thought in my head more than a few seconds."

Shaking her head, Missy sighed.

"My trying to recall anything isn't gong very well. I think I was given another shot, but I'm not sure. Everything now, is a blur. Maybe I will remember more in the days to come. I didn't really learn anything about what was happening in Nefasylum. I just knew I would probably be killed."

Reaching out, Luke patted Missy's leg.
"You did fine. Whatever was in those shots, would warp your thinking anyway. I had several in my

time at that place. It seems the people working for whoever runs Nefasylum are all cut from the same cloth. This guy that gave you the shots, and questioned you, is just another Doctor Neff. A man with no scruples, no humanity, and probably paid well to do what he does."

Hank nodded. "And for now, that guy, and his henchman are sitting in jail. I plan on writing a series of articles exposing that place and those behind it."
He turned to Missy.
"I just want you to feel better so you can help me with that. I've already had a few calls from more people who have stories to tell."

Missy stared at him.
"Already? That's a good sign. I'll be back in the office in the morning."

Hank laughed, shaking his head.
"Take a day or two for that. In fact, it will probably take several days to get all that clown make-up off your face."

Missy laughed. "A couple days and this swelling will be gone, and I won't need it."
Then she frowned.
"Do you know anything about why those other two were in the basement?"

Hank shook his head.
"Not yet, but I'm going to talk with Spencer tomorrow, and see what he's learned."

Luke was nodding. "Good idea."
Then he frowned.
"I wonder if that security guard who opened the door knows he's out of a job?"

Hank held up his paper.
"If he reads the paper, he'll know tomorrow."

Chapter 13

After Hank talked with Spencer, he asked Luke if he and Missy could meet at Luke's house. When Luke agreed. Hank called Missy to tell her.
"I told Luke this time we'd meet after dinner, so no one had to feed the others."

Missy had laughed at that, and said she would be there for sure. They had agreed on seven thirty. Hank had arrived before Missy, but only a few minutes. When she arrived, Luke had her follow him to the kitchen.
"Take a seat. Do you want some coffee, or hot chocolate? I have both ready."

Missy smiled. "I'll take some hot chocolate. I can get my own though."

Luke shook his head at her.
"Just sit down. My kitchen, I do the serving."

Laughing, Missy sat down, and turned to Hank. She pointed at her face.
"Look, no make-up. Not for a clown, or otherwise."

Taking a good look at Missy's face, Hank nodded.
"The swelling is almost gone. Faster than I thought it would be. I have to say there's still some

bruising. Is this your way of showing me you're ready to come back to work?"

Before she answered, Missy took the cup Luke was handing her. Thanking him, she turned and nodded at Hank.
"I'm more than ready. I wasn't made for just sitting at home. I need to fill my mind with work."

As Luke took his seat, Hank nodded. He looked from one face to the other.
"So, I imagine you're both anxious to hear if I learned anything from Spencer."
When the two nodded, Hank continued.
"Spencer was able to talk with the two who were abducted. The man who was in the basement is Frank Harmon. Frank is a hypnotherapist who lives down in Colorado. He was actually doing work for the government. He used his skills to help military personnel who were suffering with PTSD. He was leaving his office one night when he was abducted. That was six months ago."

Both Luke and Missy let out a gasp at the information. Missy frowned.
"Six months? That poor guy. Why would they do that? If he was doing what sounds like important and helpful work, and for the government, why abduct him?"

Luke answered the question.

"Someone who knows about hypnotherapy would be useful to the types of people who ran Nefasylum. If you could hypnotize a soldier to do something they wouldn't normally do, you have a unique weapon."

Hank nodded. "That's my thinking as well. It's also something many think has happened numerous times in the past. From what Spencer said, Frank was held somewhere in Colorado until the last month. That was when he was brought to Nefasylum. Whatever was being planned, was getting set up to happen soon. Arrangements are being made for Frank to be taken back to his home. He will also be given police protection."

Turning to Missy, Hank sighed.

"That reminds me. Spencer said to let you know, he was having you, and your apartment, placed under surveillance. Either Spencer, or one of his deputies, have already started driving by your place at night. That's for your protection. Spencer arrested those two men, but there are others who wouldn't like you talking about what happened."

A wide eyed stare, was followed by Missy shaking her head.

"I never thought about that. I hate having them need to do that, but I should be thankful they are. You're

right about there being others like those two that abducted me."
Then she frowned.
"What about the woman? Who is she? Another hypnotherapist?"

Shaking his head, Hank grunted out a laugh.
"Not quite. Her name is Mara Stevens. She's from Nevada. You might be surprised to learn this, I know I was. Mara's job there is as a psychic. Mara was on a trip to this area, when she was taken. She was in that basement almost a week."

Luke stared at Hank.
"A psychic? I don't get it."

Hank sighed. "It is a bit strange. I think they felt Mara could teach soldiers how to read minds, or see the future. or something stranger. I can't really tell you what they might have been thinking. She'll be going into protective custody, down in Nevada. These people, the ones who are connected to Nefasylum, they don't care about destroying lives, using people, or the future harm they cause. Whatever their disgusting agenda is, they will stop at nothing to achieve it. That's why the series of stories we put in the paper will be so important. They can help people heal, and hopefully stop those still trying to do these atrocious things in the present, and in the future."

Missy nodded. "Right, we can't let people forget, not just what happened in the past, but what could. Like you said, occur in the future."
She frowned, looking at Hank.
"Have you learned anymore about that company? Those who own RavenWing should be held responsible as well."

Hank shrugged, "Not a whole lot, but I'm still digging. I'll be glad when I get my number one reporter back to help."

Unsure she had heard correctly, Missy stared wide eyed at Hank.
"You mean it? You're going to move me into an actual reporter position?"

Hank nodded. "You deserve it. We both know George is retiring. I'm already interviewing for a new receptionist. I mean if you want the job. I can try and find someone else if you don't."

Reaching out, Missy slapped Hank's shoulder. "Don't you dare. I'm the reporter for you."

Luke was smiling. "Congratulations, Missy. I hope moving up in the newspaper world won't stop you from visiting with an old man."

Leaning over, Missy kissed Luke's cheek.
"Nothing could stop me from doing that."

The three replaced the conversation of Nefasylum with talk that was more optimistic. They all knew in the days to come, they would learn more than they ever thought possible. Things that they wished had never happened. Stories from people whose lives had been horribly altered by Doctor Neff in that run down building on the hill. Families, who years later, never set foot in the place, but still felt the repercussions from those who did.

The one thing they all hoped was to have the building torn down, and healing to begin.
If the latter was even possible.

Chapter 14

Fall became winter. The holidays seemed to fly by. Then Spring came, bringing with it, the images of life beginning anew. Missy, Hank, and Luke, hoped the optimism that came with the green leaves, and budding flowers, would be something felt in other parts of their lives.

Luke was enjoying his new friendships with Hank, and Missy. He spent a lot of time with Hank out in his garage, as the weather became warmer. The two men had found they had much more in common than just the asylum, and what had happened there.

They were in the garage when Luke handed Hank a paper. Staring at it, Hank frowned.
"What's this?"

Luke smiled. "Oh just a little thing they call a home title. I don't have any family to speak of. You've told me on more than one occasion how much you like this place, and have since you were a kid. I know you are renting your place, but have thought about buying a home. Now, you won't have to worry about doing that. I don't know how many years I'll walk this earth, but this place is yours. You could move in now, or wait until I'm gone. I'd like to see this house in the hands of someone who appreciates it. I hope that's okay with you."

Hank stared at the man.
"It's more than alright with me. That's too much to give someone though."

Shaking his head, Luke laughed.
"Let's just say you having this place, I feel I can rest in peace and call it good."

Hank smiled. "I don't know what to say. Thanks hardly seems to be enough."

Smiling back, Luke placed his hand on Hank's shoulder.
"It's more than enough. Let's get back to work on this truck."

For Missy, Spring did mean new beginnings.
She had the dream job of being a reporter.
On top of that, in the days when Spencer, and his deputies watched her apartment, and her, Missy had come to know Spencer better than she had before. The two had begun dating. They were even talking about marriage, maybe kids. Missy felt content. Though her nights were often filled with nightmares of her time at Nefasylum, she was healing.

She was in her office at the newspaper building when Hank stepped over frowning.
"What are you doing? It doesn't look like you're busy writing articles."

A sigh, was followed by a grin.
"Actually, I was trying to answer a few of these e-mails. Now that you've made me a reporter, and even added my name to the by-line of any articles concerning Nefasylum, I've been getting dozens, and dozens of job offers. They've been coming in from all over the country."

Hank frowned. "Wait a minute, your answering job offers on your e-mail while on company time?"

Missy laughed. "Don't worry, I'm turning them all down. We're finding less and less to report on Nefasylum. Hell, the building will be bulldozed next week, so I need something to pass my time."

Hank rolled his eyes. "We could always put you back out at the front desk, taking calls."

Missy grinned, then stared at Hank, blue eyes narrowing.
"Actually, I was also checking on something else. Have you heard much about the people getting sick, a few even dying, up at the chemical plant in Gainesville? It's only about twenty miles from here.

I could take a drive up there and see what's happening."

Rolling his eyes, Hank shook his head.
"Oh no you don't. Not on your own anyway. You'd think you'd have learned your lesson down in that basement at Nefasylum."
He let out a sigh.
"Let me grab a jacket, It's still a bit chilly outside. We can drive up in my SUV. I trust my driving more than yours."

Laughing, Missy stood, and nodded.
"I'll just grab my jacket too. Thanks Hank."

I want to thank all who picked up this book. I hope you enjoyed the story. I am fascinated by the things you hear about what the Government, and their secret operations might be doing. Most things people hear about are those that come from the past. You have to wonder if they continue to this day, and will go one in the future. The only way to know is to watch, and report. I hope that continues.

Thanks to my family, and friends who have followed me on the journey. You have supported me in the madness, and I am forever grateful.
As I continue to try and get to that goal of publishing 100 books since 2012, I hope you will all stay with me on the journey.
Life is good, and even thoughts of the life beyond this world, feel like an experience I'd love to be still writing about. Maybe, I can be a ghost writer when that time comes.
I always did believe anything is possible.
A bit about the other books is on the pages ahead if you'd like to know about them. You can find them all on Amazon, or Barnes and Noble. Just look for P.S. Winn, that's me. Thanks again to all.

P.S. Winn

The Novels

"Foretold" - Predictions come true as the ultimate showdown between good and evil begins. One place in Idaho is a shining light for those looking for hope.

"Voices" - A serial killer is murdering the down and out as the voices whisper. Sadly, even the killer can't remember the murders.

"Obligations" - A car wreck takes Josh to the other side and then back as he is given the obligation to find and destroy the evil that followed.

"Capernicious" - Sue and Cheryl have been friends a long time, now Sue is married and finds her husband changing after starting his new job at Capernicious academy. The two friends have to get together to figure out why.

"B.A. 47" - An explosion next to a subdivision reveals a conspiracy and the witnesses in the neighborhood are in a race for their lives.

"Pacific Passage" - Winning a dream cruise, Bree takes her two best friends along. The nightmare begins when the cruise wrecks as it goes through a vortex.

"Suppression" - The actual true story of a man with a device that can end pollution and create cheap or free energy, guess who doesn't want that? Maybe it could be the government and their big oil backers.

"Lies in Shadows" - Lucretia was sheltered all her life, now with her parents dead, and a memory block crumbling, she is finding the horrific reasons why.

"Phases" - George is a nice guy, well, except that one night a month. Usually, he'd give you the shirt off his hairy back. Hopefully Jake can help George out.

"Mystic Valley" - If the government hid the paraphernalia from Area 51, you might find it in Mystic Valley"

The New Moon Killer" - Not all killers strike on the full moon. Becca and Tom are hunting one in the present while Becca heads to past lives through hypnosis to help with Chronic Pain

"Healings" - As a healer, Andrew has walked the earth since Biblical times. Meeting his soul mate might change all that.

"Superstition Canyon" - The Native Americans hid a relic years ago, now it has been found and the finder has to be stopped before he unleashes something terrible.

"Collisions" - Using an old Ouija Board, Lindy causes a rip in the veil between worlds, now they are colliding.

"Viewings" - As a remote viewer, Greg has seen a lot of strange things, but what he finds in an old barn will amaze even him.

"At Hidden Lake" - Something is hidden at the lake, three friends are in danger when they uncover the hidden secrets from WW II and who was behind them.

"A Gradual Decline" - Serial killer Riley Jackson is going to be executed. Rick Holton has the exclusive story, but does he have too much empathy for the killer?

"Judgments" - A serial killer is committing gruesome crimes, who, how and why, may shock you.

"Of Jeebies and Journeys" - The Jeebies have stolen something from Ellie. Her husband Jed is going to journey to the other side to try and find her and give it back.

"Into The Doorways" - Terrene is a parallel world in trouble, they have opened the doorways to try and find a new home. Not as easy as it sounds.

"Correlations" - Lacey is going to find everything is related and will need help from this world and the next to stop the past that is haunting her.

"Whitmore Hills" - The psychiatric center houses those with mental problem, but they aren't the crazy ones.

"Transpirations" - Jess is a healer, but a fringe religious group is going to try and stop that as a conspiracy unfolds.

"Realities" - Sometimes the realities of life are not what we expect and may be more supernatural than real.

"Presence" - Andi Moore has a special gift, she can see spirits. As woman are abducted, Andi is asked to help. This is only the second time in her life she has come face to face with evil.

"Just Divine" - When two friends buy a coffee shop with a house attached, they are going to find a twisted mystery with a paranormal twist.

"Disturbance & Destruction" - The worst has happened, a man brought the world to destruction. Now it is 2047 and a few survivors that escaped to a parallel world are headed back to search for any survivors and to see if anything of the old world is left.

"Deadly Afflictions" - Dealing with anxiety, Amy wants to change her life, instead she lands in the middle of murder, mystery and more.

"Fatal Pretense" - The people of Aurora were excited to have much needed jobs when a drug company opened its' doors. Then strange happenings and deaths have them scrambling to find the truth, before more bodies pile up.

"Believing" - In a world of good and bad, some people are gathering and discovering they have special talents and gifts that are a bit supernatural and badly needed to make the world a better place for all.

"Unveilings" - Coming back from a near death experience and waking in the hospital was a miracle. When Tina realizes she also came back without the veil that hides this world from the next, her life changes forever.

"Scrutiny" – A missing doctor, concerned friends and a hospital up to some strange and illegal things. Who's watching who in this medical conspiracy thriller?

"Perdish Inn" – Two friends step into horror, supernatural and suspense when they step into Perdish Inn. Stepping out is going to be the hardest thing.

"Power and Pretense" – When the wrong man is elected, nothing feels right. When he belittles the intelligence agencies, including the FBI, how are they supposed to protect the man? Death threats against him, mean they have to.

"House of Lies" - After being laid off, Ray Larsen finds a dream job. He is ecstatic, that is, until things happen and make all he hoped for a nightmare.

"The Earth Watcher" - Before Covid-19 hit and then ran rampant around the world, people were watching earth. The story follows one of the watchers to find why they have decided to hold a vigil over this earth.

"Persistence" - Years from now, many things have changed, but a few haven't. New viruses come and go. Families change because of circumstances. Ethan and Abby are an example of that.

"Entries" - Nick Porter inherited his grandparents' small farm, he also inherited the troubles he has found under a shed on the property.

"Disruptions" - In life things happen that can change a person's perspective. Can the way people handle things, good and bad, change all they know and how they act?

"Plots and Premises" - Trying to grab the American dream to own a home, a place for family, The Warner's land instead in a deadly conspiracy. Based on a true story.

"Midnight On Owl Mountain" - Owl Mountain has also been a strange place, full of secrets. Lights above the mountain during the four solstices of the year add to the mystery. Don and Kyle are heading up, despite the danger to see if they can unlock the mysteries.

"Between" - Jack is a special person, a near death experience takes him to a place between worlds. After he is sent back, Jack has a special job that has him interacting with spirits between worlds.

"Obligations Return" - Eight years ago, Josh died. He traveled to the other side and was sent back with an Obligation. His job was t find the evil that followed and destroy it somehow. He thought he was done with evil, but no, it is back.

"Mammoth Cave" - Three friends know all about the horrors that have happened up at Mammoth cave. The stories go back to when Native Americans were driven from the area. But what is gossip and what is truth? And who will be the next to die?

"Pine View Institute" - When Kelly wakes, she finds herself in strange surroundings. Her memory is gone. When she meets Gina, the two women find that Pine View Institute is the place where nightmares begin.

"Strategies" - Kim knew the U.S. Supreme Court sent the abortion issue back to the states. She didn't think much about that. Not until she ended up raped and pregnant. Trying to decide what to do, Kim finds conspiracy, hatred, and that many choose money over people.

"The Vale" - A google search sends Cade on an adventure. A recent disappearance of a town, people and all, is a mystery Cade has to look into. Heading to the area in Montana, where it happened, Cade finds more than he bargained for. Science fiction blends with science fact in a strange adventure.

"Burdens" - Life hands us all kinds of burdens to carry. When a company does the unthinkable, the burdens for Fran Goldsmith and her family, are unbearable.

"Overt" - A man has been murdered in the most overstated, unthinkable way. A small town deputy is doing his best to solve the mystery. Finding a killer, may be the hardest job Tony Warren ever had to deal with.

"Haven Point" - The worst has happened, it hit all over the world. So many are dead. The survivors, have to find a way to continue on, in a new, strange, world.

"75 Hawthorne Lane" - Getting a new job, and a new home, Jan felt life was like living a dream. Sadly, strange happenings are making it all more of a nightmare.

"1410" - Two friends in their sixties, had something strange happen when they were kids. They are still looking for answers to the strange phenomenon. Some people want to stop them though.

"The Intake Center" – The place is open. People who want to help are waiting. Some who land here are prepared to take the next step in the journey. Others though, need some help before they make an attempt moving through the veil.

"Channels" - Where are the children? The searches have been fruitless. Can a mind, with a special gift, see where they are? Will they be found alive and well?

"Time Points" – Do people watch over the world? Can they interfere to stop a catastrophe, or are their abilities limited by time and space?

"History of Heritage House" – What if walls could talk and tell the story of a house. What would we learn? For Ryan Marshall, a man with a special gift or two, that scenario is possible.

"The Anomaly Project" – Nell was the first to see the anomaly. Other scientists learn of the thing. They know it is headed for Earth, and would destroy life as we know it. The race is on.

"Faerilon" – Other worlds, even if they look like a fairy tale can be real. The world of Faerilon is in trouble. The people there remember a girl they called Princess Avery. They are hoping she, now called Nina. will return to help.

"Nefasylum" – People had heard the stories, and rumors of the horrific things that happened at the asylum. When the doors closed, they thought the horror was over.
They thought wrong.

Jen's Journey's 3 book series

"Perspectives" Jen's Journeys – Book 1 – Jen is a widow who feels her life has come to a stalemate. She is looking for more and buying an RV, heads into adventure she never expected.

"A Time of Thorns and Roses"
Jen's Journeys, book 2 – Jen is back and once again the older woman is in big trouble. All Jen wanted was to try and begin a new life after her husband died. That isn't happening.

"Animalistic Behavior"
Jen's Journeys Book 3 – Jen is headed to a zoo camping trip and taking her two grandsons along. Big problem when a serial killer is on the loose.

Jen's Journeys – A Triple Treat. - Now, all of Jen's journeys can be enjoyed in one collection. The series can also be read as stand alone books. The choice is the readers to make.

"Tunnels" – A three book series.

Book one – "Avowals"
Nikki is called home when her father becomes sick. It's not until she gets there that she finds he caught the disease through the tunnels in a parallel world.

Book Two – "Panacea"
The search is on. The tunnels are open, and Nikki along with a special team of people, are stepping into the unknown.

Book Three – "Empyrean"
Amazing worlds open for Nikki and the team. Some are splinter worlds, others are similar to earth, but have seen awful things. In some worlds the team find horror, happiness, and things they never thought possible.

Tunnels – The Complete Saga.
All three books of the tunnels series in one collection. Grab a big book for an epic journey.

Young Adult Short Stories adults can enjoy

"Stretched Stories" - Tall tales that can be enjoyed by all ages and great to share.

"Stretched Stories 2" - More tall tales, read them aloud with the ones you love.

Young Adult Series, adults can enjoy

"Parallel Adventures 1 – Into the Caves" - When twins Jayden and Jenny step into the caves near their home, they discover the doorway to parallel worlds.

"Parallel Adventures 2 – Secrets Revealed" - The Parallel adventures continue. Jenny and Jayden are going to find amazing secrets about the past, present and future.

"Parallel Adventures 3 - Strange Happenings" - More adventures as the twins find the strange things going on and the people they are helping are themselves.

"Parallel Adventures 4 – Spooks and Spirits" - The twins are twenty and embarking on strange adventure as they find themselves headed into hell and back to maintain balance in all the worlds.

"Parallel Complete" - Now readers, young and old, can grab all four books in the parallel adventures series.

Short Story Collections

"Visitations" - Can others reach us after they pass on to the other side, you might be surprised.

"Heartfelts" - Life hands us hard things, poems and short stories tackle them with uplifting messages.

"Wings to Whispers" - A touch or a whisper may just be a loved one saying hello from the other side.

"Life Bridges" - In life we build bridges, three people fought prejudice to build the bridges of understanding that touch our lives.

"Tidbits and Treasures" - Poems, poems and more poems, plus a few insights added in.

"Hypnagogia" - Short stories that delve into the strange place between being awake and being asleep. It is also the place my stories actually come from.

"Legends of Lore" - Short stories that examine some folklore, a few angels, and a bit more.

Comic books for adults

"The Golden Years" - Humorous comics about growing old.

"The Golden Years 2" - More comics, no one wants to grow old, don't go gracefully, have fun.

Books for Children

"The Alphabet Book" - Preschool reader, learn the alphabet with happy rhymes

"The Number Book" - Preschool rhymes teach numbers with fun

"The Secret Life of Goats" - Fun early reader, rhyming tale about a family of goats and the secrets they keep.

"No, Jimmy, No" - Nobody wants to hear the word "No". Jimmy looks for a place without the word. Fun rhyming early reader.

www.ingramcontent.com/pod-product-compliance
Lightning Source LLC
Chambersburg PA
CBHW052249220526
45471CB00001B/251